Submerged Rice

Morphological, Molecular and Genetic Analyses

Pranab Basuchaudhuri
Formerly Senior Scientist
Indian Council of Agricultural Research
New Delhi, India

CRC Press is an imprint of the
Taylor & Francis Group, an **informa** business
A SCIENCE PUBLISHERS BOOK

First edition published 2024
by CRC Press
2385 NW Executive Center Drive, Suite 320, Boca Raton FL 33431

and by CRC Press
4 Park Square, Milton Park, Abingdon, Oxon, OX14 4RN

CRC Press is an imprint of Taylor & Francis Group, LLC

Library of Congress Cataloging-in-Publication Data (applied for)

ISBN: 978-1-032-66053-0 (hbk)
ISBN: 978-1-032-66340-1 (pbk)
ISBN: 978-1-032-66343-2 (ebk)

DOI: 10.1201/9781032663432

Typeset in Palatino Linotype
by Prime Publishing Services

Dedication

To
My beloved
Little Granddaughter
Parnika Basuchaudhuri

Preface

In our childhood (50's) we have seen more than 50 percent of rice growing areas were inundated by water during monsoon in West Bengal, Odisha and Bihar in eastern region of India. The rice varieties grown were tall and good bunch of straw used for cattle feed and thaching huts. With time because urbanization, dams, roads and better removal or discharge of water from basin area water stagnation reduced appreciably. Other reason is introduction of dwarf high yielding varieties with less straw and heavy panicle with full of grains. During 70's this is termed as "Green Revolution". However, so much of land in the world under stagnant or water logged condition, which requires special attention. Fortunately, because of the situation for a long time number of site specific well adopted varieties became in vogue with medium grain production and sustainable in the agroclimatic conditions.

Ever increasing population and food security concern had urged scientists to find a way out so that rice grain production in these area is appreciably enhanced. In view of that International Rice Research Institute, Malina, Phillippines and leading Universities of various countries take an endeavor to improve some high yielding dwarf rice varieties genetically that they can sustain different depth and duration of waterlogging at different stages of growth with high grain yield. After long researches they identified an indica variety FR13A, having a Chrosome 9 gene described as Sub1.

Hence Sub1 is incorporated in high yielding dwarf varieties like Swarna Sub1 and others. In India Swarna Sub1 is brought into cultivation. Now Sub1 Mega Varieties are also available for cultivation. In different countries varieties with Sub1 gene are being identified and incorporated in suitable varieties for more grain production and adaptability. It is expected in future some unique varieties will be available for cultivation.

The book is written lucidly with full of scientific informations for postgraduate students and researchers so that they will be interested in the subject. There are seven chapters in the book describing all facets of the submerged rice including flood and its historical nature, present prevailing situations, loss in terms of money, loss of rice production, effects on society and others. Details of scientific knowledge of changes and mechanism in the plant systems in respect of morphology, biochemical molecular and genetic aspects of rice with recent concepts. Lastly, agronomical advancements and yield and yield –gap analysis is also elaborated.

I will be happy if the reader drop their suggestions for betterment and book is liked and appreciated.

I express my sincere thanks to my family members for their support during the preparation of the book.

Dr. Pranab Basuchaudhuri
Hong Kong

Contents

1

Introduction

Rice (*Oryza sativa* L.) is grown across the globe and consumed by approximately 3 billion people or around 50% of the world population (Zhang et al., 2014; Zhao et al., 2020). Rice was grown on 162 million hectares and its global production was 755 million tons in 2019 (http:// www.fao.org/ t/en, accessed on 29 June 2021). The world population may rise anywhere from 9.7 to 11 billion in 2050 (https://population.un.org/ wpp/, accessed on 29 June, 2021); thus, a significant increase in rice yield will be required to feed the growing population. The global demand for rice is estimated to increase by 50% by 2050 (Ray et al., 2013). However, climate change is a major limiting factor in crop production and increases in temperature are leading to more frequent and severe drought spells and soil salinization (Budak et al., 2015).

Rice (*Oryza sativa*) is one of the world's most important food crops, and is comprised largely of *japonica* and *indica* subspecies. Researchers reconstruct the history of rice dispersal in Asia using whole-genome sequences of more than 1,400 landraces, coupled with geographic, environmental, archaeobotanical and paleo-climate data. Originating around 9,000 years ago in the Yangtze Valley, rice diversified into temperate and tropical *japonica* rice during a global cooling event about 4,200 years ago. Soon after, tropical *japonica* rice reached Southeast Asia, where it rapidly diversified, starting about 2,500 year BP. The history of *indica* rice dispersal appears more complicated, moving into China around 2,000 year BP. Researchers also identify extrinsic factors that influence genome diversity, with the temperature being a leading abiotic factor. Reconstructing the dispersal history of rice and its climatic correlates may help identify genetic adaptations associated with the spread of a key domesticated species (Gutaker et al., 2020).

1.1 Rice: Agro-environment

Rice, grown in different agro-environment, is classified generally as,

1. Lowland, rain-fed
2. Lowland, irrigated
3. Deepwater
4. Coastal wetland
5. Upland rice

There is very little opportunity to increase the area earmarked for rice and further crop intensification is constrained by limited supply of water. Therefore, the increase in supply must mainly be met by increasing crop yield through better crop, nutrient, pest and water management and the use of germplasm and biotechnological engineering with a higher yield potential. In the case of grain crops, an excellent process had been introduced to obtain healthy plants with enriched calorie and nutrient quality of grain. Also, Integrated Nutrient Management (INM), Integrated Pest Management (IPM) and numerous improved irrigation techniques are being adopted along with application of fertilizers. Genetic enrichment and modification are being employed and tested seriously (Fairhurst and Dobermann, 2002).

Rice is an important crop that serves as a staple food for over 3.5 billion of the world's population and also serves as food security for many countries in Africa and Asia (Chukwu, 2019). It is widely grown in varied environmental conditions, ranging from sea-coasts to high altitudes. The major rice-growing areas are greatly affected by floods, caused by river discharge, excessive rain-water accumulation, and tidal movements. Globally, one-third of the rice-cultivated areas are deep-water and rain-fed lowland ecosystems, which account for about 50 million hectares (Bailey-Serres and Voesenek, 2008). These areas are prone to frequent flooding, which could be attributed to poor drainage systems of excessive rain-water during the rainy season. Rice has anaerobic tolerance; however, extreme flooding, either complete or partial submergence, may result in various environmental stressors. Hence, rice cultivars respond differently to the variations in flood-water regime. Therefore, it can be concluded that a flood-tolerant cultivar at one location cannot be transferred to another location (Dar et al., 2017). There are three categories of flooding that occur at different crop-growth stages and varied durations, which are typical of rice growing agroecosystems (Fig. 1.1).

China is one of the largest rice-producing countries in the world, producing more than 30% of the world's rice output (Liu et al., 2019). Approximately 65% of the Chinese people consume rice as their staple food (Huang et al., 2014). In most rice-planted areas of China, single-

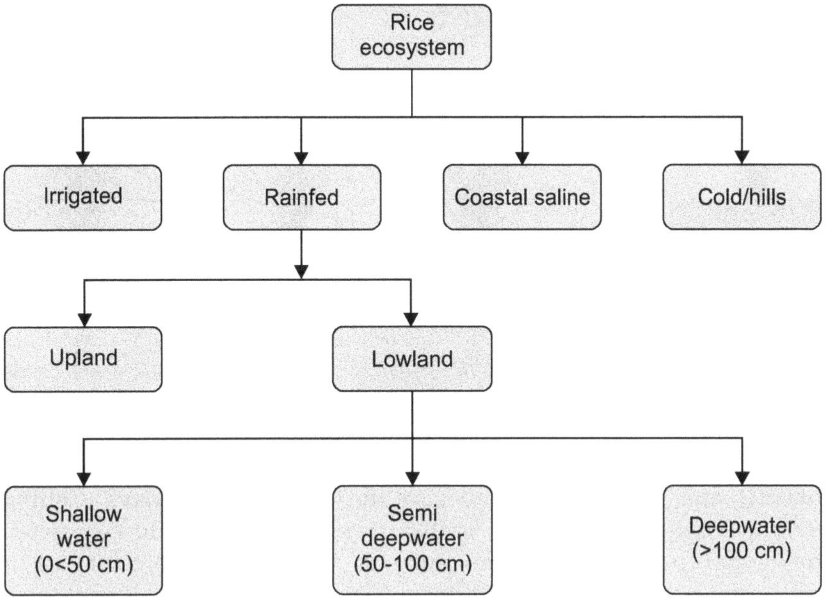

season rice is usually seeded in May and transplanted in June. However, the frequent heavy rainfall events occurring from May to June often flood rice fields in the lowland areas where rice plants are at the vegetative growth stage (Hao, 2012). Floods occurring at this stage negatively affect rice growth and grain yield by reducing the number of tillers, plant leaf area and dry weight above ground, and in the worst case, by causing high mortality (Ella et al., 2011; Kato et al., 2014; Huang et al., 2020). A previous study showed that in China more than four million hectares of rice fields are prone to flooding and the submergence caused yield loss of more than one million tons every year (Chen et al., 2005).

Submergence affects crops in low-lying areas. Submergence can be defined on the basis of duration, depth, and frequency. Flooding causes submergence and damage to rice crops. Two types of environment cause submergence: flash floods and deepwater. Flash flood submergence is defined by water levels rising rapidly and plants remaining submerged for one to two weeks. Deepwater submergence is defined by water depths greater than 100 cm persisting for months.

There is a large variation in the depth, duration and frequency of submergence which are dictated by the local topography. Therefore, rice plants should have adaptability to a particular submergence environment. One common type of submergence environment is the flash flooding previously mentioned. This is defined as a rapid surge of flooding that

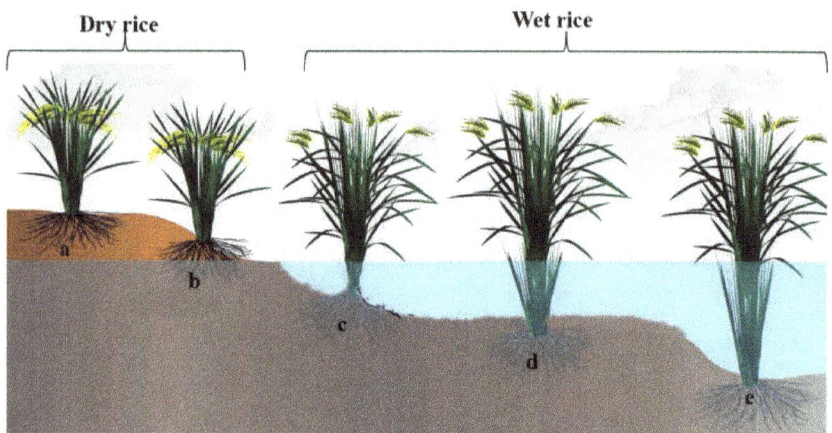

Fig. 1.1: Agro-ecosystem of rice culture.

subsides after several days and lasts no longer than 10 days (Catling, 1992). Most rice cultivars die within several days of being completely submerged, but some cultivars, such as FR 13A, are more tolerant of submergence (Mazaredo and Vergara, 1982).

1.2 Some Statistics of Flood Damage

The agro-ecosystem is significantly affected by flooding, which has been on continuous intensification for the past few decades (Ismail et al., 2013). Flooding is undoubtedly the third most vital constraint for achieving high productivity and after heat and drought, seriously affects crop production. Seasonal and unseasonal crop damage due to the occurrence of severe flooding amounts to billions of dollars in yield losses annually. For example, as reported by Bailey-Serres et al. (2012), in the United States, a 12-year study on losses of crop production revealed that flooding is the second threat after drought. It was reported that these two abiotic stresses (flooding and drought) accounted for a 70% drop in harvest, in 2011 alone. In that year, there were insurance pay outs of over 3 billion US$ due to flooding, with over 1.6 billion paid on soybean and maize. In Pakistan, over 4.45 billion US$ worth was lost due to flooding in rice, cotton, and wheat in 2010 (Ahmad, 2013). Similarly, an increase in summer rainfall causes water stagnation and flooding, with economic consequences across Europe (Olesen, 2011). At the moment, one of the most flood-threatened crops is rice. About 30% (700 million) of people living in abject poverty (i.e., daily income less than 1$) in Asian countries reside in flood-prone rice-cultivating regions of South Asia, with Nepal, Bangladesh, and India accounting for half of the above-stated figure. In Nepal, 15% of the total cultivated area of 1.5 million hectare is affected by

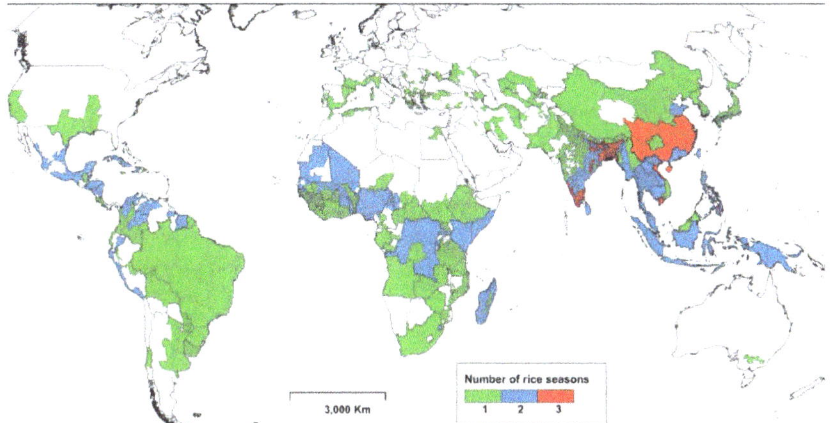

Fig. 1.2: Global rice ecosystem with number of rice seasons.

floods every year (ABPD, 2013). Similarly, out of 16.1 million hectares of rice-growing areas in India, 5.2 million are occasionally affected by floods, while in Bangladesh, 1.6 million hectares of rice fields are periodically affected by floods, out of a total area of 2.65 million hectares dedicated to rice cultivation (Bailey-Serres et al., 2010). Other most vulnerable deltas include Ayeyerwaddy delta in Myanmar, Red River and Mekong deltas in Vietnam. According to Wassmann et al. (2009), these deltas provide up to 70% of total rice cultivation areas in these countries, and continuous flooding greatly threatens their food security. In fact, over 35% of rice-growing areas, mostly in Africa and Asia, where food insecurity is predominant are prone to flooding (Bailey-Serres et al., 2012). Excessive flooding portends a major risk to human livelihood and is also a major contributor to vulnerability and poverty in marginalized rural population. It was projected that flood-prone coastal region will experience an increase in rainfall pattern, especially in tropical and subtropical regions (Sarkar et al., 2014). The increase in flooding regime associated with climate change and the necessity to step up the current agricultural yield potential by 70% remains a great challenge to feed the human population, which is expected to reach over nine billion by 2050 (Godfray et al., 2010). Therefore, the development of resilient crops to combat flooding is required to nip this problem in the bud (Fig. 1.2).

1.3 Wetland Rice System

Wetland rice systems in Asia make a major contribution to global rice supply. The system is also able to maintain soil fertility on a sustainable basis. The essential components of wetland rice culture comprise cultivation of land in the wet or flooded state (puddling), transplanting of rice seedlings into puddled rice paddies, and growing the rice crop under flooding. The land is dry or flood-fallowed during the turnaround period between two crops. Following these cultural practices, two or three crops of rice or rice with upland crops in sequence are grown. However, in the present context of increasing freshwater scarcity, there is a case to shift from the traditional method of growing rice to ways that are water-wise. In this context, it is crucial that the benefits of the wetland rice system on soil fertility and productivity are considered. This article examines the benefits of growing rice in flooded conditions on soil fertility and its maintenance. Research has shown that the wetland rice system (growing rice in submerged soils) has a great ameliorative effect on chemical fertility: largely by bringing pH in the neutral range, resulting in better availability of plant nutrients and accumulation of organic matter. The benefits of growing rice using submerged conditions must be considered and weighed in the context of a likely shift to growing rice with water-management practices that are water-wise.

Plants are completely flooded, ranging from days to weeks before the flooding recedes. Singh et al. (2017) defined flooding tolerance in this case as the plant's ability to survive 10 to 14 days of complete submergence and continue their growth process after the flood recedes, with little damage to plant morphology. This problem is endemic in river basin areas, and often leads to complete yield loss due to flash flood. Xu et al. (2006) cloned the SUB1 gene from the FR13A, 4 of 16 landrace from India, which was adapted to submerge at the vegetative stage. The third category of flooding occurs in lowland rice cultivation, where the water level ranges from 20 to 50 cm depending on whether it is a rain-fed or irrigated system. However, due to seasonal changes and rain duration, the water level might exceed 50 cm, up to 4 m, which lasts for a prolonged period of time and could partially or completely submerge the plant, depending on the growth stage. This type of flooding may remain for a few months depending on the area. Many landraces, also known as floating rice, are adapted to this type of flooding and can extend up to 25 cm per day as the flood level increases (Singh et al., 2017). Generally, stagnant flooding occurs as a result of heavy rain, which can be deep-water flooding or flash flooding. Under deep-water flooding, plants increase their stem length (internode) to escape the submergence. However, after the flood subsides, affected plants are prone to lodging and eventually this leads to death due to exhaustion of reserved nutrients within a few days. Contrarily, during flash floods, plant growth is restricted and resumes after occurrence of the flood (Winkel et al., 2013). Regardless of the temperature, duration, and depth of the flood, flood has a major impact and threatens plants with shortage of carbohydrates and cellular energy, which hinder growth and development (Fig. 1.2).

Rice is basically an autogamous plant propagating mainly through seeds produced by self-pollination. Normally anthesis starts shortly after the panicle emerges from the flag leaf. The opening of flowers within a panicle starts from top to bottom, taking nearly one week for the anthesis to complete. The spikelets generally open from morning 9–10 a.m. but this varies with the prevailing weather conditions, particularly temperature and humidity. After the anthesis, the protruded anthers wither away. Each anther consists of more than 1,000 pollen grains, which disperse immediately after the spikelet opens and germinates on the surface of stigma in about 3 minutes after pollination. Only one pollen tube reaches the ovule in about an hour after pollination, to initiate double fertilization and triple fusion. The maturation of anther and ovule occurs simultaneously without any time gap within a spikelet. The ability to germinate pollen lasts only for a few minutes under favourable temperature and moisture conditions whereas ovules can remain viable for several days after maturation. The pollen viability of cultivated species lasts only for 3–5 minutes while that of wild species it is up to 9 minutes (Oka and Morishima, 1967; Koga et al., 1971). Even though rice is basically a self-pollinated crop, natural outcrossing can occur up to 0.5 to 4% (Oka, 1988). Most of the wild species have a longer stigma, which protrudes well outside the spikelet, thereby facilitating increased percentage of out crossing (Virmani and Edwards, 1983). Different ecotypes of *O. sativa, indica* have higher percentage of out crossing than *japonica* (Oka, 1988). Although *O. sativa* is an annual crop propagated by seed, it can be maintained vegetatively as a perennial crop. The perennial character of *O. sativa* might have been inherited from its ancestor *O. rufipogan*, a perennial species (Morishima et al., 1963). Under natural conditions after harvesting the tiller rice (*Oryza sativa* L.), 85 buds on the basal nodes start growing again. These new tillers are called *ratoon* and can be grown best under long day conditions (Fig. 1.3).

The plant height varies according to variety and environmental conditions, ranging from 0.4 m in dwarf varieties to 5 m in some deep-water floating rice.

Generally taller varieties have a greater penetration of roots than shorter varieties. Similarly varieties with good tillering habit have a well-developed root system. The direct seeded crop develops deeper but has a poorly developed root system, whereas the transplanted crop has a shallow but well-developed root system. In case of deep-water rice varieties, the nodal roots develop even on higher nodes as the plant elongates.

The stem above the ground level consists of solid nodes and hollow internodes and is commonly called a 'culm'. Based upon the habitat and species, the elongation of internodes varies. Rapid elongation is observed mostly after the emergence of panicle from the flag leaf. The lowermost buds on the crowded nodes, just at the ground level or below that, develop

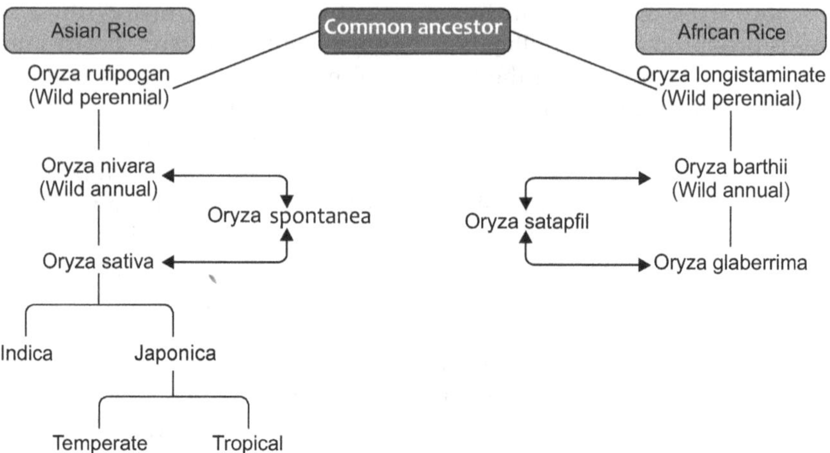

Fig. 1.3: Schematic representation of evolutionary pathways of Asian and African cultivated rice.

into tillers. The tillering habit depends on varieties, spacing, fertilizers, and cultural operations.

1.4 Rice: Cultivated species *Oryza sativa* L.

The genus *Oryza* has 25 species out of which only two are cultivated species. In the Indian sub-continent, *Oryza sativa* is the cultivated species. India is rich in rice diversity by virtue of being a part of primary centre of origin on one hand and having many secondary centres, on the other. Thousands of land races and hundreds of improved cultivars adapted to different agro-ecological conditions are present in India. According to an estimate, 75,000 to 1,00,000 land races are present in India (Randhawa et al., 2006) and in the last four decades (1965–2007), more than 836 high-yielding rice varieties were released (Shobha Rani et al., 2008). India is also known for quality rice, like Basmati and other fine-grain aromatic types and many cultivars with medicinal value, viz. Gauthuhan, Njavara, Alcha, Laicha, etc. More than 90,000 accessions have been imported and made available to researchers and farmers in India (Singh et al., 2001). Introductions from IRRI during the 1960s have led to the green revolution. Characterization of germplasm based on identifiable stable morpho-physiological traits has enabled their use for economic importance. Characterization of wild rice species has revealed their importance as potential donors for various biotic and abiotic stresses. A number of elite lines and cultivars characterized with resistance to biotic and abiotic stresses and quality characters were developed through conventional breeding, using the available genetic variability in the germplasm of *O. sativa* to the maximum extent. However, genetic variability for different traits, like resistance to stem borer, sheath blight, tungro virus, etc., is limited in the cultivated germplasm. Therefore, for further improvement, the cultivated rice gene pool has been widened by introgression of genes from wild species for tolerance/resistance to major biotic and abiotic stresses as well as for quality characters using different newly-developed techniques.

Submergence variety are FR 13A, FR 43B, Chakia59, CN540, S22, Madhukar, S24, S25, S28, Boku, Boyan, Bayyahunda, HbDW8, Solpona, Sail badal, Dhola badal, Kolasali, Boga bordhan, Khajara, Dhusara, Nali Baunsagaja, Rongasali, O. rufipogon, Sub-1.

Waterlogging types are Jhingasail, Kalakhersail, Patnai 23, NC496, Tilakkachari.

Deep-water kind are HBJ1, Nageri Baoorsa, Kekoa bao, Jalmagna, Jaladhi 1, Jaladhi 2.

Among the above transgenics, only a few have completed different steps of development and come up to field level whereas others are still in developmental phases. Golden rice and Bt-rice are a few examples which are on the verge of release.

The seed yield depends on the variety yield potential, crop management, plant protection measures, soil fertility level and finally on the prevailing climatic conditions. In general, the high-yielding varieties in fertile delta soils yield a maximum up to 7.0 to 7.5 t/ha with an average of 5.0 to 6.0 t/ha. In rain-fed uplands, the maximum seed yield is 3.0 to 3.5 t/ha with an average of 2.0 t/ha.

Submergence can occur at any time, starting from seed-germination to the maturity stage. The submergence tolerance gene *SUB1A* improves submergence tolerance at the vegetative stage (Neeraja et al., 2007) but not during the germination (Sarkar, 2012) and reproductive stages (Ray et al., 2016a). *SUB1* imparts tolerance by suppressing elongation, whereas, during germination, greater elongation of coleoptiles is required for survival and crop establishment under water. During the reproductive stage, protection of reproduction organs is more important than quiescence of elongation growth. Submergence causes mechanical injury to the developing reproductive organs of the plants. 'Baliadhan', a traditional rice cultivar with *SUB1 QTL* tolerated the reproductive stage submergence better compared to 'Swarna-Sub1' and 'Swarna' (Ray et al., 2016a). Can the cultivar with *SUB1* tolerate SF or not? It was observed that a cultivar with an inherent short height, like Swarna-Sub1, showed intolerant reactions under medium depth SF (Vergara et al., 2014). However, a cultivar with long stature like FR13A tolerates the SF easily. It shows that tolerance to SF is independent of *SUB1* (Singh et al., 2011). The elongation of shoots is common, both in GSOD and SF tolerances. It is still unknown whether or not the same genes are expressed in both cases, or whether they govern the tolerance or not? However, cultivars tolerant to both GSOD and SF are available. So, the development of cultivars which can tolerate flooding at different growth stages is possible. Combinations of different plant parameters can encounter the various flood-related stresses efficiently.

1.5 Salinity and Stagnant Flooding/Submergence

Flooding with salty water due to climate change and other associated events poses a great threat to rice grown in coastal plains (Wassmann et al., 2004). Salinity reduces plant growth and plants become weak. The already weak rice plants die out early compared to healthy plants under submergence (Sarkar and Ray, 2016). Salt-tolerant plants maintain their health better under salt-stress compared to the intolerant ones, so improving salt-tolerance is imperative for rice grown on coastal plains and even to counteract the deleterious effects of flooding.

Besides genetic factors, environmental factors, such as light, turbidity, concentrations of oxygen and carbon dioxide, and even the concentration of ethylene in flood waters, influence survival under submergence (Setter et al., 1997; Panda et al., 2006; Das et al., 2009). However, the scenario between submergence under saline (12 dS m^{-1}) and non-saline water (0.17 dS m^{-1}) was not so different; FR13A, tolerant to submergence but susceptible to salinity, behaved similarly under both the situations (Sarkar and Ray, 2016). In contrast to submergence with the combined effects of salinity and partial submergence, even the submergence-tolerant cultivar FR13A shows greater injury. Cultivars tolerant to the combined stresses of salinity and partial submergence preserve more pigment, maintain better chloroplast structural and functional ability (Zheng et al., 2009), improve oxygen transport through formation of aerenchyma tissues and maintain better antioxidant systems (Haddadi et al., 2016). Notwithstanding its agricultural and biological importance under changing climates, the regulatory mechanisms of adaptation to saline submergence and SF have rarely been studied in rice.

The seed yield depends upon the variety yield potential, crop management, plant protection measures, soil fertility level and finally on the prevailing climatic conditions. In general, the high-yielding varieties in fertile delta soils yield maximum up to 7.0 to 7.5 t/ha with an average of 5.0 to 6.0 t/ha. In rain-fed uplands, the maximum seed yield is 3.0 to 3.5 t/ha with an average of 2.0 t/ha.

References

Agri-Business Promotion and Statistics Division (ABPD), *Statistical Information on Nepalese Agriculture 2012/2013*; Agriculture Business Promotion and Statistics DivisionMoADC, HMG/N, Singh Durbar: Kathmandu, Nepal, 2012.

Ahmad, Z. (2013). Disaster risks and disaster management policies and practices in Pakistan: A critical analysis of Disaster Management Act 2010 of Pakistan. *Int. J. Dis. Risk Reduct.*, 4: 15–20.

Bailey-Serres, J., and Voesenek, L.A.C.J. (2008). Flooding stress: Acclimations and genetic diversity. *Annu. Rev. Plant Biol.*, 59: 313–339.

Bailey-Serres, J., Fukao, T., Ronald, P.C., Ismail, A.M., Heuer, S., and Mackill, D. (2010). Submergence tolerant rice: SUB1's journey from landrace to modern cultivar. *Rice*, 3: 138–147.

Bailey-Serres, J., Lee, S.C., and Brinton, E. (2012). Waterproofing crops: Effective flooding survival strategies. *Plant Physiol.*, 160: 1698–1709.

Budak, H., Hussain, B., Khan, Z., Ozturk, N.Z., and Ullah, N. (2015). From genetics to functional genomics: Improvement in drought signaling and tolerance in wheat, *Front. Plant Sci.*, 6: 1–13.

Catling, D. (1992). *Rice in Deep Water*, London: MacMillan Press.

Chen, Y., Yan, Q., and Xiao, G. (2005). Progress in research of submergence tolerance in rice. *Chinese Agricultural Science Bulletin*, 21(12): 151–153 (In Chinese with English abstract).

Chukwu, S.C., Rafii, M.Y., Ramlee, S.I., Ismail, S.I., Oladosu, Y., Okporie, E. et al. (2019). Marker-assisted selection and gene pyramiding for resistance to bacterial leaf blight disease of rice (*Oryza sativa* L.). *Biotechnol. Biotechnol. Equip.*, 33: 440–455.

Dar, M.H., Chakravorty, R., Waza, S.A., Sharma, M., Zaidi, N.W., Singh, A.N. et al. (2017). Transforming rice cultivation in flood prone coastal Odisha to ensure food and economic security. *Food Secur.*, 9: 711–722.

Das, K.K., Panda, D., Sarkar, R.K., Reddy, J.N., and Ismail, A.M. (2009). Submergence tolerance in relation to variable floodwater conditions in rice. *Environ. Exp. Bot.*, 66: 425–434.

Ella, E.S., Dionisio-Sese, M.L., and Ismail, A.M. (2011). Application of silica before sowing negatively affects growth and survival of rice following submergence. *Philippine Journal of Crop Science*, 36(2): 1–11.

Fairhurst, T.H., and Dobermann, A. (2002). Rice in the Global Food Supply. *Better Crops International*, Vol. 16.

Godfray, H.C.J., Beddington, J.R., Crute, I.R., Haddad, L., Lawrence, D., Muir, J.F., and Toulmin, C. (2010). Food security: The challenge of feeding 9 billion people, *Science*, 327: 812–818.

Gutaker, R.M., Groen, S.C., Bellis, E.S. et al. (2020). Genomic history and ecology of the geographic spread of rice. *Nat. Plants*, 6: 492–502.

Haddadi, B.S., Hassanpour, H., and Niknam, V. (2016). Effect of salinity and waterlogging on growth, anatomical and antioxidative responses in *Mentha aquatica* L. *Acta Physiol. Plant*, 38: 119–129.

Hao, Y. (2012). The utilization of tolerance rice germplasm and the influence of submergence on carbohydrate accumulation and transformation, Master's Thesis, College of Plant Science & Technology, Huazhong Agricultural University, Wuhan, China (in Chinese with English abstract).

Huang, S., Wang, L., Liu, L., Fu, Q., and Zhu, D. (2014). Non-chemical pest control in China rice: A review. *Agronomy for Sustainable Development*, 34: 275–291.

Huang, S., Jia, Y., Liu, P., Dong, H., and Tang, X. (2020). Effect of ultrasonic seed treatment on rice seedlings under waterlogging stress. *Chilean Journal of Agricultural Research*, 80: 561–571.

Ismail, A.M., Singh, U.S., Singh, S., Dar, M.H., and Mackill, D.J. (2013). The contribution of submergence–tolerant (Sub1) rice varieties to food security in flood–prone rain-fed lowland areas in Asia. *Field Crops Res.*, 152: 83–93.

Kato, Y., Collard, B.C.Y., Septiningsih, E.M., and Ismail, A.M. (2014). Physiological analyses of traits associated with tolerance of long-term partial submergence in rice, *AoB Plants*, 6: Plu058.

Koga, Y., Akihama, T., Fujimaki, H., and Yokoo, M. (1971). Studies on the longevity of pollen grains of rice *Oryza sativa* L. I. Morphological changes of pollen grains after shedding, *Cytologia*, 36: 104–110.

Liu, Z., Zhang, W., Zhang, Y., Chen, T., Shao, S., Li, Z. et al. (2019). Assuring food safety and traceability of polished rice from different production regions in China and Southeast Asia using chemometric models. *Food Control*, 99: 1–10.

Mazaredo, A.M., and Vergera, B.S. (1982). Physiological differences in rice varieties tolerant of and susceptible to complete submergence. pp. 327–342. *In:* 1981 International Deep Water Rice Workshop. International Rice Research Institute, Los Banos, Philippines.

Morishima, H., Hinata, K., and Oka, H.I. (1963). Comparison of modes of evolution of cultivated rice species, *Oryza sativa* L., *O. glabberima Steud. Japan J. Breed.*, 12: 153–165.

Neeraja, C.N., Maghirang-Rodriguez, R., Pamplona, A., Heuer, S., Collard, B.C.Y., Septiningsih, E.M. et al. (2007). A marker-assisted backcross approach for developing submergence-tolerant rice cultivars. *Theor Appl. Genet.*, 115: 767–776.

Oka, H.I., and Morishima, H. (1967). Variation in breeding systems of wild, perennis rice. *Oryza perennis, Evolution*, 21: 249–258.

Oka, H.I. (1988). *Origin of Cultivated Rice.* Japan Sci. Soc. Press, Tokyo, pp. 245.

Olesen, J.E., Trnka, M., Kersebaum, K.C., Skjelvåg, A.O., Seguin, B., Peltonen–Sainio, P. et al. (2011). Impacts and adaptation of European crop production systems to climate change. *Eur. J. Agron.*, 34: 96–112.

Panda, D., Rao, D.N., Sharma, S.G., Strasser, R.J., and Sarkar, R.K. (2006). Submergence effects on rice genotypes during seedling stage: Probing of submergence driven changes of photosystem 2 by chlorophyll a fluorescence induction O-J-I-P transients. *Photosynthetica*, 44: 69–75.

Randhawa, G.J., Bhalla, S., Chalam, V.C., Tyagi, V., Verma, D.D., and Hota, M. (2006). *Document on Biology of Rice* (*Oryza sativa* L.) *in India.* National Bureau of Plant Genetic Resources, New Delhi and Project Coordinating and Monitoring Unit, Ministry of Agriculture.

Ray, D.K., Mueller, N.D., West, P.C., and Foley, J.A. (2013). Yield trends are insufficient to double global crop production by 2050. *PLoS ONE*, 8: e66428.

Ray, A., Panda, D., and Sarkar, R.K. (2016). Can rice cultivar with submergence tolerant quantitative trait locus (SUB1) manage submergence stress better during reproductive stage? *Archives of Agronomy and Soil Science*, Doi: 10.1080/03650340.2016.1254773.

Sarkar, R.K., and Bhattacharjee, B. (2012). Rice genotypes with SUB1 QTL differ in submergence tolerance, elongation ability during submergence and re-generation growth at re-emergence. *Rice* (N.Y.) 5:7 doi.10.1007/s 12284-011-9065-z.

Sarkar, R.K., Das, K.K., Panda, D., Reddy, J.N., Patnaik, S.S.C., Patra, B.C. et al. (2014). *Submergence Tolerance in Rice: Biophysical Constraints, Physiological Basis and Identification of Donors.* Central Rice Research Institute: Cuttack, India, pp. 1–36.

Sarkar, R.K., and Ray, A. (2016). Rice withstands complete submergence even in saline water: Probing through chlorophyll a fluorescence induction O-J-I-P transients. *Photosynthetica*, 54: 10.1007/s11099-016-0082-4.

Setter, T.L., Ellis, M., Laureles, C.V., Ella, E.S., Senadhira, D., Mishra, S.B. et al. (1997). Physiology and genetics of submergence tolerance in rice. *Ann Bot.*, 79: 67–77.

Shobha Rani, N., Prasad, G.S.V., Subba Rao, L.V., Sudharshan, I., Manish, K. Pandey, Babu, V.R., et al. (2008). High-yielding rice varieties in India, *Technical Bulletin No. 33.* Directorate of Rice Research, Rajendranagar, Hyderabad 500 030, Andhra Pradesh, India, pp. 188.

Singh, R.V., Deep Chand, Brahmi, P., Verma, N., Tyagi, V., and Singh, S.P. (2001). Germplasm exchange. pp. 69–89. *In:* Dhillon, B.S., Varaprasad, K.S., Singh, M., Archak, S., Srivastava, U., and Sharma, G.D. (Eds.). *National Bureau of Plant Genetic Resources: A Compendium of Achievements,* National Bureau of Plant Genetic Resources, New Delhi, India.

Singh,S., Mackill, D.J., and Ismail, A.M. (2011). Tolerance of longer-term partial stagnant flooding is independent of the SUB1 locus in rice. *Field Crops Research,* 121: 311–323.

Singh, A., Septiningsih, E.M., Balyan, H.S., Singh, N.K., and Rai, V. (2017). Genetics, physiological mechanisms and breeding of flood-tolerant rice (*Oryza sativa* L.). *Plant Cell Physiol.*, 58: 185–197.

Vergara, B.S., and Mazaredo, A. (1975). *Screening for Resistance to Submergence under Greenhouse Conditions.* Bangladesh International Rice Research Institute: Dhaka, Bangladesh, pp. 67–70.

Vergara, G.V., Nugraha, Y., Esguerra, M.Q., and Ismail, A.M. (2014). Variation in tolerance of rice to long-term stagnant flooding that submerges most of the shoot will aid in breeding tolerant cultivars. *AoB Plants,* 6:plu055.

Virmani, S.S., and Edwards, J.B. (1983). Current status and future prospects for breeding hybrid rice and wheat. *Adv Agron.*, 36: 145–214.

Wassmann, R., Hien, N.X., Hoanh, C.T., and Tuong, T.P. (2004).Sea level rise affecting the Vietnamese Mekong Delta:water elevation in the flood season and implications for production. *Clim.Change,* 66: 89–107.

Wassmann, R., Jagadish, S.V.K., Sumfleth, K., Pathak, H., Howell, G., Ismail, A. et al. (2009). Regional vulnerability of climate change impacts on Asian rice production and scope for adaptation. *Adv. Agron.*, 102: 91–133.

Winkel, A., Colmer, T.D., Ismail, A.M., and Pedersen, O. (2013). Internal aeration of paddy field rice (*Oryza sativa*) during complete submergence–importance of light and floodwater O_2. *New Phytol.*, 197: 1193–1203.

Xu, K., Xu, X., Fukao, T., Canlas, P., Maghirang-Rodriguez, R., Heuer, S. et al. (2006). Sub1A is an ethylene-response-factor-like gene that confers submergence tolerance to rice. *Nature,* 442: 705.

Zhang, Q., Chen, Q., Wang, S., Hong, Y., and Wang, Z. (2014). Rice and cold stress: Methods for its evaluation and summary of cold tolerance-related quantitative trait loci. *Rice,* 7: 24.

Zhao, M., Lin, Y., and Chen, H. (2020). Improving nutritional quality of rice for human health. *Theor. Appl. Genet.*, 133: 1397–1413.

Zheng, C., Jiang, D., Liu, F., Dai, T., Jing, Q., and Cao, W. (2009). Effects of salt and waterlogging stresses and their combination on leaf photosynthesis, chloroplast ATP synthesis, and antioxidant capacity in wheat. *Plant Sci.*, 176: 575–582.

2

History of Flood

2.1 Introduction

Floods are among the most powerful forces on Earth. Human societies worldwide have lived and died with floods from the very beginning, spawning a prominent role for floods within legends, religions, and history. Inspired by such accounts, geologists, hydrologists, and historians have studied the role of floods on humanity and its supporting ecosystems, resulting in a new appreciation for the many-faceted role of floods in shaping our world. Part of this appreciation stems from ongoing analysis of long-term stream-flow measurements, such as those recorded by the U.S. Geological Survey's (USGS) stream-flow gauging network. But the recognition of the important role of flood in shaping our cultural and physical landscape also owes to increased understanding of the variety of mechanisms that cause floods and how the types and magnitudes of floods can vary with time and space. The USGS has contributed to this understanding through more than a century of diverse research activities on many aspects of floods, including their causes, effects, and hazards. This summarizes a facet of this research by describing the causes and magnitudes of the world's largest floods, including those measured and described by modern methods in historic times, as well as floods of prehistoric times, for which the only records are those left by the floods themselves.

In general, large river basins produce large floods, but large unit discharges in the moist tropics can result in floods of disproportionately large size. The largest meteorologic floods from river basins are larger than about 500,000 sq. kms. This broad framework for the temporal and spatial distribution of floods is but a launching point for ongoing USGS flood research. Together with many cooperating State, local, and federal agencies, academic institutions and international partners, USGS studies focused on floods are taking many forms. These activities include developing new ways to measure and monitor floods and to map flood hazards, investigating the complex relations between floods and persistent climatic patterns, such as the El Niño Southern Oscillation and the Pacific Decadal Oscillation, and understanding the role of flood processes in forming and maintaining landscapes and ecologic systems. All of these activities have the goal of providing high-quality scientific information that can be used to make informed decisions regarding the management of floods and their benefits and hazards throughout the world.

But ambiguities remain regarding the mechanisms at play. With the dominant flood season differing region-wise (winter for northern and north-western Europe, but summer for central Europe, for example), the winter-time NAO can explain only part of this story. The related role of prevailing temperatures in the observed flood-rich periods also remains an open question. Earth-system modeling, twinned with process-based hydrological models could be used to capture the complexity of natural and human processes on the ground that enhance or suppress flooding. It could also provide further insight into the extent to which the association of cooler-than-usual conditions with most flood-rich periods is causative or correlative, and how much the most recent flood-rich period is a result of human-induced climate warming. Blöschl et al. (2020) clearly demonstrate the potential of historical climatology, but there are yet more avenues by which to advance this approach to reconstruct past climate. Written evidence of climate can become particularly discontinuous in the deep past, but tree-ring evidence, for example (measurements of tree-ring widths, densities and isotopic composition, which correlate with climate conditions) is more continuous over long periods. However, tree-ring evidence usually reflects only the growing-season climate, whereas written evidence can attest to weather for all seasons (Table 2.1).

Floods are the most common (and among the most deadly) natural disasters in the United States. They have brought destruction to every state and nearly every county, and in many areas they are getting worse. As global warming continues to exacerbate sea-level rise and extreme weather, our nation's floodplains are expected to grow by approximately 45 per cent by century's end. Here's how climate change plays a role in flooding, and how we can better keep our heads above water.

A flood is the accumulation of water over normally dry land. It is caused by the overflow of inland waters (like rivers and streams) or tidal waters, or by an unusual accumulation of water from sources such as heavy rains or dam or levee breaches.

Table 2.1: Quaternary floods with discharges greater than 100,000 cu. m. per second.

Basin number	River basin[a]	Country	Basin area (10^3 km^2)[b]	Station	Station area (10^3 km^2)	Station latitude (degrees)	Station longitude (degrees)	Peak discharge (m^3/s)	Date	Flood type
1	Amazon	Brazil	5,854	Obidos	4,640	1.9S	55.5W	370,000	June 1953	Rainfall
2	Nile	Egypt	3,826	Aswan	1,500	24.1N	32.9E	13,200	Sept. 25, 1878	Rainfall
3	Congo	Zarie	3,699	Brazzaville B.	3,475	4.3S	15.4E	76,900	Dec. 27, 1961	Rainfall
4	Mississippi[c]	USA	3,203	Arkansas City	2,928	33.6N	91.2W	70,000	May 1927	Rainfall
5	Amur	Russia	2,903	Komsomolsk	1,730	50.6N	138.1E	38,900	Sept. 20, 1959	Rainfall
6	Parana	Argentina	2,661	Corrientes	1,950	27.5S	58.9W	43,070	June 5, 1095	Rainfall
7	Yenisey	Russia	2,582	Yeniseysk	1,400	58.5N	92.1E	57,400	May 18, 1937	Snowmelt
8	Ob-Irtysh	Russia	2,570	Salekhard	2,430	66.6N	66.5E	44,800	Aug. 10, 1979	Snowmelt
9	Lena	Russia	2,418	Kasur	2,430	70.7N	127.7E	189,000	June 8, 1967	Snowmelt/ Ice Jam
10	Niger	Niger	2,240	Lokoja	1,080	7.8N	6.8E	27,140	Feb. 1, 1970	Rainfall
11	Zambezi	Mozambique	1,989	Tete	940	16.2S	33.6E	17,000	May 11, 1905	Rainfall
12	Yangtze	China	1,794	Yichang	1,010	30.7N	111.2E	110,000	July 20, 1870	Rainfall
13	Mackenzie	Canada	1,713	Norman wells	150	65.3N	126.9W	30,300	MAy 25, 1975	Snowmelt
14	Chari	Chad	1,572	N'Djamena	600	12.1N	15.0E	5,160	Nov. 9, 1961	Rainfall
15	Volga	Russia	1,463	Volgograd	1,350	48.5N	44.7E	51,900	MAy 27, 1926	Snowmelt
16	St. Lawrence	Canada	1,267	La Salle	960	45.4N	73.6W	14,870	May 13, 1943	Snowmelt
17	Indus	Pakistan	1,143	Kotri	945	25.3N	68.3E	33,280	1976	Rain/ Snowmelt
18	Syr Darya	Kazakhstan	1,070	Tymen'-Aryk	219	44.1N	67.0E	2,730	June 30, 1934	Rain/ Snowmelt
19	Orinoco	Venezuela	1,039	Puente Angoatura	836	8.1N	64.4W	98,120	Mar. 6, 1905	Rainfall

20	Murray	Australia	1,032	Morgan	1,000	34.0S	139.7E	3,940	Sept. 5, 1956	Rainfall
21	Ganges	Bangladesh	976	Harding Bridge	950	23.1N	89.0E	74,060	Aug. 21, 1973	Rain/ Snowmelt
22	Shatt al Arab	Iraq	967	Hit (Euphrates)	264	34.0N	42.8	7,366	May 13, 1969	Rain/ Snowmelt
23	Orange	South Africa	944	Buchuberg	343	29.0S	22.2E	16,230	1843	Rainfall
24	Huanghe	China	894	Shanxian	688	34.8N	111.2E	36,000	Jan. 17, 1905	Rainfall
25	Yukon	USA	852	Pilot Station	831	61.9N	162.9W	30,300	May 27, 1991	Snowmelt
26	Senegal	Senegal	847	Bakel	218	14.9N	12.5W	9,340	Sept. 15, 1906	Rainfall
27	Colorado[c]	USA	808	Yuma	629	32.7N	114.6W	7,080	Jan. 22, 1916	Rainfall
28	rio Grande[c]	USA	805	Roma	431	26.4N	99.0W	17,850	1865	Rain/ Snowmelt
29	Danube	Romania	788	Orsova	575	44.7N	22.4E	15,900	April 17, 1895	Snowmelt
30	Mekong	Vietnam	774	Kratie	646	12.5N	106.0E	66,700	Sept. 3, 1939	Rainfall
31	Tocantins	Brazil	769	Itupiranga	728	5.1S	49.4W	38,780	April 2, 1974	Rainfall
32	Columbia[c]	USA	724	The Dalles	614	45.6N	121.2W	35,100	June 6, 1894	Snowmelt
33	Daling	Australia	650	Menindee	570	32.4S	142.5E	2,840	June 1890	Rainfall
34	Brahmaputra[d]	Bangladesh	650	Bahadurabad	636	25.2N	89.7E	81,000	Aug. 6, 1974	Rain/ Snowmelt
35	Sao Francisco	Brazil	615	Traipu	623	9.6S	37.0W	15,890	April 1, 1960	Rainfall
36	Amu Darya	Kazakhstan	612	Chatly	450	42.3N	59.7E	6,900	July 27, 1958	Rain/ Snowmelt
37	Dnieper	Ukraine	509	Kiev	328	50.5N	30.5E	23,100	May 2, 1931	Snowmelt

a Basin larger than 500,000 square kilometers for which reliable data were not available include the Nelson River in North America; the Jubba, Irharhar, Araye, Tafassasset and Qattar Rivers in Africa: and the Kolyma and Tarim Rivers in Asia.

2.1.1 River Flooding

This occurs when a river or stream overflows its natural banks and inundates normally the dry land. Most common in late winter and early spring, river flooding can result from heavy rainfall, rapidly melting snow, or ice jams. According to one study, approximately 41 million U.S. residents are at risk from flooding along rivers and streams.

2.1.2 Coastal Flooding

More than 8.6 million Americans live in areas susceptible to coastal flooding when winds from a coastal storm, such as a hurricane or nor'easter, push a storm surge—a wall of water—from the ocean on to land. Storm surge can produce widespread devastation. There are also increasing numbers of shallow, non-life-threatening floods caused by higher sea levels; these high tide floods (also known as 'nuisance' or 'sunny-day' floods) occur when the sea washes up and over roads and into storm drains as the daily tides roll in.

2.1.3 Flash Floods

These quick-rising floods are most often caused by heavy rains over a short period (usually six hours or less). Flash floods can happen anywhere, although low-lying areas with poor drainage are particularly vulnerable. Also caused by dam or levee breaks or the sudden overflow of water due to a debris or ice jam, flash floods combine the innate hazards of a flood with speed and unpredictability and are responsible for the greatest number of flood-related fatalities.

2.1.4 Urban Flooding

Flash floods, coastal floods, and river floods can occur in urban areas, but the term 'urban flooding' refers specifically to flooding that occurs when rainfall—not an overflowing body of water—overwhelms the local storm-water drainage capacity of a densely populated area. This happens when rainfall runoff is channeled from roads, parking lots, buildings, and other impervious surfaces to storm drains and sewers that cannot handle the volume.

2.1.4.1 Flooding Causes

Many factors can go into the making of a flood. There are weather events (heavy or prolonged rains, storm surge, sudden snowmelt), and then there are the human-driven elements, including how we manage our waterways (via dams, levees, and reservoirs) and the alterations we make to land. Increased urbanization, for example, adds pavement and other impermeable surfaces, alters natural drainage systems, and often leads to more homes being built on floodplains. In cities, under-maintained infrastructure can lead to urban flooding. More and more, flooding factors are also linked to climate change.

2.1.4.2 *Climate Change and Flooding*

Connecting climate change to floods can be a tricky endeavour. Not only do myriad weather—and human-related factors play into whether or not a flood occurs, but limited data on the floods of the past make it difficult to measure them against the climate-driven trends of floods today. However, as the IPCC (Intergovernmental Panel on Climate Change) noted in its special report on extremes, it is increasingly clear that climate change 'has detectably influenced' several of the water-related variables that contribute to floods, such as rainfall and snowmelt. In other words, while our warming world may not induce floods directly, it exacerbates many of the factors that do. According to the *Climate Science Special Report* (issued as part of the *Fourth National Climate Assessment*, which reports on climate change in America), more flooding in the United States is occurring in the Mississippi river valley, midwest, and northeast, while U.S. coastal flooding has doubled in a matter of decades.

These are some of the key ways climate change increases flood risks.

2.1.4.3 *Heavy Precipitation*

A warmer atmosphere holds and subsequently dumps more water. As the country has heated up on an average of 1.8 degrees Fahrenheit since 1901, it has also become about 4 per cent wetter, with the eastern half of the United States growing soggiest. In the northeast, the most extreme storms generate approximately 27 per cent more moisture than they did a century ago. Basically, because of global warming, when it rains, it pours more. Such was the finding of a study by the National Oceanic and Atmospheric Administration (NOAA) examining the record-breaking rainfall that landed on Louisiana in 2016, causing devastating flooding. The study determined that these rains were at least 40 per cent more likely and 10 per cent more intense because of climate change.

Looking forward, heavy precipitation events are projected to increase (along with temperatures) through the 21st century, to a level from 50 per cent to as much as three times the historical average. This includes extreme weather events known as atmospheric rivers, air currents heavy with water from the tropics, which account for as much as 40 per cent of typical snowpack and annual precipitation along the west coast.

Experts predict they will intensify, bringing as much as 50 per cent more heavy rain by the end of this century.

Of course, heavier rainfall does not automatically lead to floods, but it increases the potential for them. Even moderate amounts of rainfall can cause serious damage, particularly in places where urban flooding is on the rise.

Meanwhile, in regions where seasonal snowmelt plays a significant role in annual runoff, hotter temperatures can trigger more rain-on-snow events, with warm rains inducing faster and often earlier melting. This phenomenon is playing out in western United States, where, according to the IPCC, snowmelt-fed rivers, at least since 1950, have reached peak flow earlier in springtime. The combination of rain and melting snow can aggravate spring flooding as winter and spring soils are typically high in moisture and often still frozen, and therefore less able to absorb snow and rain runoff. Regions with higher rain-to-snow ratios, such as the northwest, are expected to see higher stream-flow—and higher flood risks.

Climate change is increasing the frequency of strong storms, a trend expected to continue through this century. In the Atlantic basin, an 80 per cent increase in the frequency of category 4 and 5 hurricanes (the most destructive) is expected over the next 80 years. And stronger storms bring greater rains. Indeed, 2017's Hurricane Harvey, which made landfall as a category 4 storm and soaked some 200,000 Houston homes and businesses with catastrophic floods, was the nation's wettest storm in nearly 70 years. It was also slow and therefore able to dump more—a result of weakened atmospheric currents from a warmer atmosphere. Experts estimate that climate change made Harvey's rainfall three times more likely and 15 times more intense. In 2018, Harvey was followed by Hurricane Florence (the second-wettest storm in nearly 70 years), which set at least 28 flood records in the Carolinas, according to the United States Geological Survey (USGS). Hurricane Maria, which hit Puerto Rico, Dominica, and the U.S. Virgin Islands in 2017, produced the most rainfall in the area of any weather event since 1956.

Even rain storms are predicted for the future, with tomorrow's hurricanes expected to be as much as 37 per cent wetter near their centre and about 20 per cent wetter as much as 60 miles away.

Stronger storms can also produce gustier winds that whip up greater storm surge, which starts as much as 8 ins. higher than a century ago because of sea-level rise. It was Hurricane Katrina's 28-foot storm surge that overwhelmed the levees around New Orleans in 2005 and caused a vast majority of deaths. And it was the combination of storm surge and high tide that led to Hurricane Sandy's inundation of coastal New York and New Jersey in 2012—flooding that could be as much as 17 times more frequent in the area's coastal regions by 2100, according to one study. Storm surge and winds can also increase the destructiveness of waves, causing them to get bigger and penetrate further inland. According to the Federal Emergency Management Agency (FEMA), waves of just 1.5 feet have been seen to cause significant damage to coastal structures.

2.1.4.4 High Seas

As ocean temperatures rise and the world's glaciers and ice sheets melt (phenomena exacerbated by climate change), global sea levels are rising. Our oceans are approximately seven to eight inches higher than they were in 1900 (with about three of those inches added since 1993 alone)—a rate of rise per century greater than for any other century in at least the past 2,000 years. And while the IPCC predicts seas around the world will rise anywhere from one foot to more than four feet above 2000 levels by century's end, NOAA's projections show that, due to regional factors, such as currents bringing water to coastlines, places such as the east coast could see seas as much as 9.8 feet higher by 2100. (Check out NOAA's interactive map that demonstrates where flooding will occur as sea levels rise).

In addition to amplifying storm surge because the water starts at a higher level, sea-level rise increases high-tide flooding, which has doubled in the United States over the past 30 years and is expected to rapidly worsen in the coming decades. According to the *Fourth National Climate Assessment*, for example, by 2045, Charleston, South Carolina could see as many as 180 tidal floods per year, compared with just 11 in 2014.

Patterns of past precipitation are inherently more spatially variable, and hence, harder to reconstruct than are those of temperature. Reconstructing flooding is even more challenging because flooding depends not only on precipitation, but also on human landscape usage, from upstream deforestation to damming, bridging and urbanization. It is also related to prevailing temperatures, which have a seasonally and regionally varying role across Europe, influencing evaporation and soil moisture and the timing of spring snowmelt. Importantly, therefore, Blöschl and colleagues establish an association in timing between all but one flood-rich period and the prevalence of lower-than-usual average temperatures. By contrast, the most recent flood-rich period in Europe (1990 to 2016, when the available data end)—for a region stretching into western and central Europe and northern Italy, and defined by the authors as being one of the most severe—is exceptional for occurring in a warming climate.

A flood chronology for the city of Cork is presented for the period 1841–1988. The 292 floods which are reported are classified into six flood types primarily based on the relative role of rainfall and tidal conditions in each flood event. Changes over 148 years in flood frequencies and flood types are outlined and assessed, primarily with respect to the significance of the atmospheric circulation patterns and wind direction, but also with reference to the occurrence of river discharge levels and tidal surges. The major cause of flooding is shown to be excessive rainfall, although high tides are also of considerable significance, especially those accompanied by storm surges.

Floods by the Yangtze river (Chang Jiang) in central and eastern China have occurred periodically and have often caused considerable destruction of property and loss of life. Among the most recent major flood events are those of 1870, 1931, 1954, 1998, 2010, and 2020.

The Yangtze river, the longest river in Asia, is one of the world's major waterways. It originates at an elevation above 16,400 feet (5,000 metres) in the Plateau of Tibet and proceeds generally eastward along a winding course until it empties into its major delta system on the East China Sea. The primary flood region is the lower course, downstream of the Three Gorges Dam, an area in which the river flows through low-lying terrain dotted by lakes, marshes, and meandering streams. Increase in the region's population has led to efforts to control the river. The Great Jinjiang Levee, completed in 1548, was one of many barriers constructed and by late 19th century, the Yangtze could drain through only four openings on the south side of the river. Consequently, sediment was deposited only on the river bottom or in Dongting Lake, causing the flood level to rise and create a lowland on the north bank. In addition, many of the lakes that had once acted as flood controlss either were cut off from the river by levees or were converted into cropland. Deforestation further reduced the capacity of the area to handle intense rains, which created more runoff.

As a result, when the lower Yangtze basin experienced sustained heavy rains, the consequences were catastrophic. The flood of 1931 covered more than 30,000 sq. miles (77,700 sq. kms.), including the cities of Nanjing and Wuhan; it ultimately led to the deaths of an estimated 3.7 million people and left 40,000,000 more homeless. Subsequently, more-effective levees were built, but the floods of 1954 and 1998 proved highly destructive and killed some 30,000 and 3,650 people, respectively. One of the major objectives of the Three Gorges Dam project was to alleviate flooding on the lower Yangtze. The dam proved effective during the extraordinarily rainy summer of 2010 by holding back much of the resultant floodwaters and thus minimizing the impact of flooding downstream. However, the dam still had to open its floodgates to reduce the high water volume in the reservoir, and flooding and landslides in the Yangtze basin killed hundreds of people and caused extensive property damage.

According to *The Journal of Flood Risk Management*, in Thailand, various studies showed that floods cause an extreme impact on the country's economy, especially the flood in 2011. The Bangkok flood caused 41 billion US$ worth of damage. In Vietnam, floods significantly affect rice production, the economy, and people's lives (Chen, Giese, and Chen). In Torti research, studies estimate that 9.6 million people are currently affected by flooding in Southeast Asia, with 5.3 million in Thailand alone.

Referencing Mediodia, a disaster-like floods can affect the flow of goods and transfers between nations. Moreover, it might lead to long-term effects on economic growth since it destroys facilities and causes loss of lives. However, this study about eight countries in Southeast Asia found that the occurrence of floods also positively impacts economic growth. Researchers state that disaster opens the opportunity for that country to replace the obsolete technology with a more efficient one, which results in a higher rate of technological process.

A longstanding prediction for the world's most populous region is finally becoming a reality. 'There's a consistency in the models that climate change in Asia would translate into more floods, into more intense rainy seasons,' said Homero Paltan Lopez, a water expert and researcher at the University of Oxford. Such an alteration was expected to affect the vast area's seasonal monsoon, making rainfall during the wet season more concentrated and with the dry season becoming longer. That is exactly what is happening and it is devastating many lives.

A study in science journal, *Nature Communications* last year estimated that 300 m people lived in places where climate-triggered flooding would probably occur by 2050, with most vulnerable being the Asian countries, such as China, India, Bangladesh and Vietnam. A study in July in *Scientific Reports* found that while flood risk is growing globally, Asia's population density and preponderance of coastal communities mean that the majority of the world's high-risk population in the next 80 years will be on the continent. "The science is getting more and more precise," said Abhas K. Jha, with the World Bank's urban and disaster risk management programme in East Asia and the Pacific. "One thing that we know for sure is that wet places will get wetter, and dry places will get drier." In China alone, 2.7 m people have been evacuated and an estimated 63 m were impacted in 2020. A total of 53 rivers are currently at or near historic high water levels and dams in the Yangtze river basin are near or above capacity, making for worst flooding in southern China since at least 1961. Meanwhile, in South Asia, 17 m people have been affected this year and the situation is likely to get worse as heavy rainfall is predicted in many parts of Asia this season. While its numbers are not as dramatic, Japan, no stranger to natural disasters, has seen increasingly dangerous weather. Record rains in Kumamoto Prefecture on the island of Kyushu killed at least 65 people in July. Parts of Chiba Prefecture east of Tokyo are still reeling from a huge typhoon in September last year that damaged more than 70,000 houses and knocked out electricity that led to days of blackouts, affecting tens of thousands.

While its numbers are not as dramatic, Japan, no stranger to natural disasters, has seen increasingly dangerous weather. Record rains in Kumamoto Prefecture on the island of Kyushu killed at least 65 people in July. Parts of Chiba Prefecture east of Tokyo are still reeling from a huge typhoon in September last year that damaged more than 70,000 houses and knocked out electricity that led to days of blackouts affecting tens of thousands (Fig. 2.1).

In total, the region's cities gained 200 m more residents in the 10 years from the year 2000. While that movement was most pronounced in China, increasingly it is Pakistan, Indonesia and India that are experiencing rapid urban growth. Thus, more people—and infrastructure—in high-risk regions will automatically make potential flooding more costly. 'Flood risk is also more people living in harm's way,' said Charles Iceland, director of Global Water Initiatives with The World Resources Institute's Food, Forests, Water and the Ocean programme. 'Populations are growing, people are settling and building industrial infrastructure in likely locations where flooding could occur.' Growth of cities, and the increasing number of Asians living alongside coasts or rivers, means the number of people in flood-prone areas has risen. Other human-driven changes, such as the widespread destruction for aquaculture of coastal mangroves—which are known to reduce storm surges and the intrusion of seawater inland—are caused by sinking land due to excessive groundwater discharge. And the loss of wetlands and other natural water sinks means many Asian cities are more prone to flooding even without factoring in climate change.

Currently, most global attention on climate focuses on mitigation-cutting greenhouse emissions—to reduce the long-term impact of climate change. Asia accounts for the majority of gross global carbon emissions, a proportion that is growing. 'When it comes to global mitigation, Asia is uniquely positioned given its share of the global economy, and

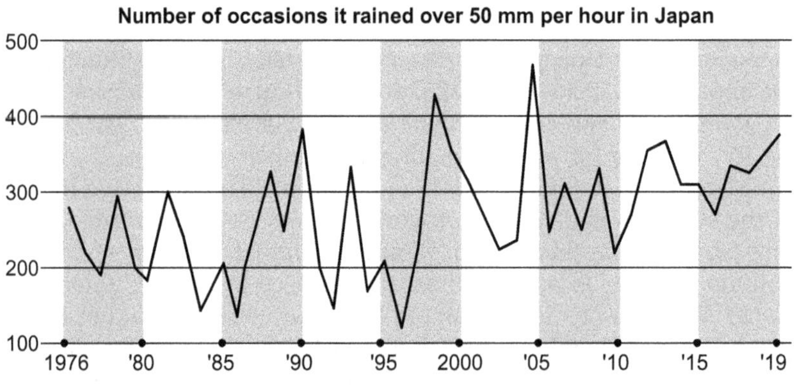

Fig. 2.1: Number of occasions of rains over 50 mm per hr.

investments in the power sector,' said Mr. Fakhrutdinov of the McKinsey Global Institute. But for floods, in the short- and medium-term mitigation has little impact as it is likely that historical emissions will result in climate change-connected intense rainfall and sea level rises—both of which make flooding more likely. There are also non-climate factors, such as migration and development, which affect the social and economic impact of floods.

Floods are common natural disasters in Asia. Flood datasets from 48 countries in Asia were collected to investigate the spatiotemporal distribution and influencing factors, using the Mann–Kendall trend test and the Spearman's rank correlation. These results show that flood occurrences and damages increased significantly in Asia, with mortality rates and deaths decreasing. Southern and eastern Asia are flood-vulnerable regions, with Central Asia being the least flood-occurrence region, and China and India are also flood-prone countries with a largest population and land area in Asia. Least flood disasters occurred in Bahrain, Cyprus, Brunei Darussalam and Singapore, with a smaller population and land area. The spatial disparities of flood disasters were positively influenced by population and land area, and negatively influenced by urbanization rate and per capita GDP. The largest proportion of flood disasters were discovered in riverine floods, followed by flash floods, with coastal floods being the least. The highest and second-highest mortality rates were observed in flash floods and coastal floods, which showed decreasing trends, and the mortality rate of riverine floods was the lowest, with an increasing trend. The rain was the main triggering origin of floods, and tropical cyclone contributed to the second, followed by snowmelt, convective storms, and dam-break flows. This analysis can help to provide a useful insight into the formulation of flood-risk maps, disaster mitigation measures, and emergency management.

'There's a consistency in the models that climate change in Asia would translate into more floods, into more intense rainy seasons,' said Homero Paltan Lopez, water expert and researcher at the University of Oxford. Such an alteration was expected to affect the vast area's seasonal monsoon, making rainfall during the wet season more concentrated, and the dry season becoming longer. That is exactly what is happening and it is devastating many lives.

The South America lands are distributed along different latitudes, from above the equator to higher than 55° S. The consequence is a vast diversity of climates due the presence of different atmospheric systems (Reboita et al., 2010). Brazil, with a large extension, shows it clearly as its lands cross low and high latitudes. It has numerous environments in each part of the country, leading to different weather conditions and precipitation patterns. As a result, the frequency of flood events is fairly high since they are distributed along the regions and also year round. According to Reboita et al. (2010), these patterns can be classified in eight different groups along South America, considering the graphs of monthly precipitation. Only the ones including Brazilian lands are going to be addressed in this work.

High rainfall and its seasonal distribution cause periodic flooding of large areas in tropical South America. Floods result from lateral overflow of streams and rivers, or from sheet-flooding by rains as a consequence of poor drainage. Depending upon the size of the catchment area, flooding can occur with one peak (e.g., in the Amazon river and its large affluents) or in many peaks (e.g., in streams and small rivers).

Vegetation cover of floodplains varies from different types of savanna and aquatic macrophyte communities to forests depending upon the hydrologic regime and local rainfall. Large differences exist in primary and secondary production due to large differences in nutrient levels in water and soils.

The higher flood events that occurred in Brazil from 1979 to 2010 were determined in order to describe the atmospheric circulation patterns associated with them. An analysis was conducted in the 23 extreme ones, according to the season (summer/winter) and the region where the event occurred. The main causes identified were the El Nino episodes, Madden-Julian Oscillation, ITCZ, and SACZ, besides smaller-scale systems. The events that were not related with these causes need further investigation (Table 2.2).

River flooding has large societal and economic impacts across Africa. Despite the importance of this topic, little is known about the main food generating mechanisms in Africa. This study is based on 13,815 food events that occurred between 1981 and 2018 in 529 catchments. These food events are classified as follows to identify the different food drivers: excess rains, long rains and short rains. Out of them, excess rains on saturated soils in western Africa, and long rains for catchments in northern and southern Africa are the two dominant mechanisms, contributing to more than 75 per cent of all food events. The aridity index is strongly related to the spatial repartition of the different food generating processes showing the climatic controls on foods. Few significant changes were detected in the relative importance of these drivers over time, but the rather short time series available prevent a robust assessment of food driver changes in most catchments. The major implication of these results is to underline the importance of soil moisture dynamics, in addition to rainfall, to analyze the evolution of food hazards in Africa.

This work provides a continental-scale overview of the food-generating mechanisms across a large sample of basins covering most regions of Africa. A classification scheme was applied to 13,815 food events that occurred in a wide variety of catchments in terms of size, topography, aridity, and land cover. The results indicated that over 75 per cent of foods are driven by excess rainfall and long rainfall

Table 2.2: Precipitation patterns in Brazil (MCC: Mesoscale Convective Complex; SACZ: South Atlantic Convergence Zone; SASA: South Atlantic Subtropical Anticyclone; LLJ: Low Level Jet; BH: Bolivian High; ITCZ: Intertropical Convergence Zone; IL: Instability Lines; CVHL: Cyclone Vortices at High Levels).

Region	Features of Annual Precipitation Cycle	Atmospheric Systems Acting
South of Brazil	Precipitation mostly homogeneous during the year with high rainfall totals.	Frontal systems and cyclones from the Pacific Ocean; subtropical CVHL; IL; subtropical MCCs; atmospheric blocking; sea breeze; SACZ; SASA; eastern Andes LLJ.
Northwest to Southwest of Brazil	Maximum totals in summer and minimum un winter.	Trade winds; easter Andes LLJ; SASA; convection due surface heating; BH; ITCZ; sea breeze; IL' tropical MCC; frontal systems; subtropical CVHL; cyclones.
North of the Northern region of Brazil and Northeast coast	Maximum rainfall in the first half of the year.	ITCZ; convection due surface heating; tropical MCC; trade winds; sea breeze; IL; tropical CVHL; SASA; frontal systems.
Northeastern Backlands of Brazil	Maximum rainfall in summer and minimum in winter with low totals.	ITCZ; tropical CVHL; frontal systems; SASA; descendent branch of zonal circulation with Amazon convection activity.

episodes. Both processes are related to soil saturation, either before or during a food event, indicating their role in triggering food events. This finding has practical consequences related to food forecasting or the analysis of the impact of climate change on foods. It is indeed necessary to distinguish the influence of soil saturation conditions in addition to that of episodes of intense rainfall. The spatial patterns suggest that climate is the main explanatory factor, with flood-generating processes strongly influenced by aridity, but also modulated by catchment properties. Two main patterns were identified: western Africa with a dominance of excess rainfall, and north/South Africa but also other semi-arid regions with a mixture of dominant processes. In these regions, one needs to be careful with food frequency analysis due to the potential presence of a mixture of distributions. Overall, no significant changes over time in the dominant food drivers across regions were detected, except a slight reduction of excess rain in West Africa linked to a decrease of antecedent soil moisture prior to foods. The results confirm to a large extent the findings obtained in other continents, indicating that soil moisture excess is the prevailing driver of flooding. Yet, the notable difference highlights that in Africa, compared to other regions, long rains are almost equivalent to the role of excess rainfall to explain the occurrence of foods, in particular for semi-arid regions that are predominant in this continent. This demonstrates

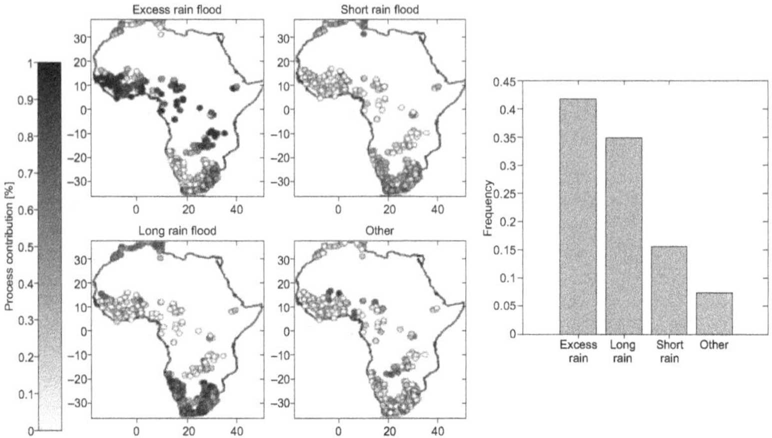

Fig. 2.2: Results of the food event driver's classification. The left panels show the relative contribution of the different food-generating processes for each basin, the histogram on the right shows the relative importance of each food-generating processes for all basins.

the importance of considering the dynamics of soil saturation at different temporal resolutions, in addition to rainfall, to better understand the food occurrence in different parts of Africa (Fig. 2.2).

A flood happens when water overflows or soaks land that is normally dry. There are a few places on Earth where people don't need to be concerned about flooding. Generally, floods take hours or even days to develop, giving residents time to prepare and evacuate. Sometimes, floods develop quickly and with little warning.

A flood can develop in a many ways. The most common is when rivers or streams overflow their banks. These floods are called riverine floods. Heavy rain, a broken dam or levee, rapid ice melt in the mountains, or even a beaver dam in a vulnerable spot can overwhelm a river and send it spreading over nearby land. The land surrounding a river is called a flood-plain.

Coastal flooding, also called estuarine flooding, happens when a large storm or tsunami causes the sea to rush inland.

Floods are the second-most widespread natural disaster on Earth, after wildfires. All 50 states of the United States are vulnerable to flooding.

2.2 Effects of Floods

When floodwaters recede, affected areas are often blanketed in silt and mud. This sediment can be full of nutrients, benefiting farmers and agribusinesses in the area. Famously fertile flood plains, like the Mississippi river valley in the American midwest, the Nile river valley in Egypt, and the Fertile Crescent in the Middle East have supported agriculture for thousands of years. Yearly flooding has left millions of tons of nutrient-rich soil behind.

However, floods have enormous destructive power. When a river overflows its banks or the sea moves inland, many structures are unable to withstand the force of water. Bridges, houses, trees, and cars can be picked up and carried off. Floods erode soil, taking it from under a building›s foundation, causing the building to crack and tumble. Severe flooding in Bangladesh in July 2007 led to more than a million homes being damaged or destroyed.

Floods can cause even more damage when the waters recede. The water and landscape can be contaminated with hazardous materials, such as sharp debris, pesticides, fuel, and untreated sewage. Potentially dangerous mould can quickly overwhelm water-soaked structures.

As flood water spreads, it carries disease. Flood victims can be left for weeks without clean water for drinking or hygiene. This can lead to outbreaks of deadly diseases, like typhoid, malaria, hepatitis A, and cholera. This happened in 2000, as hundreds of people in Mozambique fled to refugee camps after the Limpopo river flooded their homes. They soon fell ill and died from cholera, which is spread by unsanitary conditions, and malaria, spread by mosquitoes that thrived on the swollen river banks.

In the United States, floods are responsible for an average of nearly 100 deaths every year, and cause about $7.5 billion in damage.

China's Yellow river valley has seen some of the world's worst floods in the past 100 years. The 1931 Yellow river flood is one of the most devastating natural disasters ever recorded when almost a million people drowned, and even more were left homeless.

2.3 Natural Causes of Floods

Floods occur naturally. They are part of the water cycle, and the environment is adapted to flooding. Wetlands along river banks, lakes, and estuaries absorb flood waters. Wetland vegetation, such as trees, grasses, and sedges, slow the speed of flood waters and more evenly distribute their energy. According to the U.S. Environmental Protection Agency (EPA), the wetlands along the Mississippi river once stored at least 60 days of flood water (today, Mississippi wetlands store only 12 days of flood water. Most wetlands have been filled or drained).

Floods can also devastate an environment. The most vulnerable regions are those that experience frequent floods and those that have not flooded for many years. In the first case, the environment does not have time to recover between floods. In the second case, the environment may not be able to adapt to flood conditions.

In August 2010, Pakistan experienced some of the worst floods of the century. The annual monsoon, on which Pakistani farmers and consumers rely, was unusually strong. Tons of water drenched the nation. The Indus river burst its banks. Because the river flows almost directly through the narrow country, almost all of Pakistan was affected by flooding.

2.4 Flood and Rice Cultivation

Rice cultivation has long been considered to have originated from seeding of annual types of wild rice somewhere in subtropics, tropics or in the Yangtze river basin. That idea, however, contains a fatally weak point when we consider the tremendous difficulty for primitive human to seed any cereal crop in the warm and humid climate, where weed thrives all year round. Instead of the accepted theory, we have to see a reality that vegetative propagation of edible plants is a dominant form of agriculture in such regions. The possibility is discussed that Job's tears and rice, two cereal crops unique to the region, might have been developed via vegetative propagation to obtain materials for medicine or herb tea in backyard gardens prior to cereal production. This idea is supported by the fact that rice in temperate regions is still perennial in its growth habit and that such backyard gardens with transplanted taro can still be seen from the Yunnan Province of China to Laos. Thanks to a detailed survey of wild rice throughout China for 1970–1980, it is now confirmed that a set of clones of wild rice exist in shallow swamps in Jiangxi Province, an area with severe winter cold. In early summer, ancient farmers may have divided the sprouting buds and spread them by transplanting into flooded shallow marsh. Such a way of propagation might have faster improved less productive rice through a better genetic potential for response to human interference than quick fixation in seed propagation, because vegetative parts are heterogeneous. Obviously, such a primitive manner of rice cultivation did include the essential parts of rice farming, i.e., nursery bed, transplanting in flooded field of shallow marsh-like. Transfer from the primitive nursery to true nursery by seed may have later allowed rice cultivation to be extended to northern regions. In thus devised flooded cultivation, there were a series of unique advantages, i.e., continuous cropping of rice in a same plot, no soil erosion, slow decline of soil fertility, availability of minerals resulting in high yield per unit area, which have collectively encouraged highly productive cereal cultivation in the warm and humid region. Rice cultivation in marshy land is also favourable to raise fish culture, both of which constitute a nutritionally balanced base. Development of irrigation technology to construct flooded farms gave strong bases for stable rice-cultivating society, which in the end formulated the rise of ancient kingdoms of Yue and Wu in China in BC 6th–5thth centuries. They were direct descendants of those people who had developed the unique rice cultivation from the era of Hemudu culture, which is dated back to 5000 BC. Their movement to the south is considered to have established rice-cultivating communities in South China and Southeast Asia, while to the north it transferred the rice-based technology to ancient Korea and Japan, establishing there a base for a civilized society.

Rice cultivation under flooded conditions is highly sustainable. In comparison with other field crops, flooded rice fields produce more of the greenhouse gas methane but less nitrous oxide, having no-to-very-little nitrate pollution of the groundwater, and use relatively little to no herbicides. However, flooding is a constraint for rice production in years and areas of high rainfall. Although rice is well known for its cultivation in flooded soils, most cultivars cannot survive submergence for more than a week. Flash floods leading to complete submergence of rice plants for 10–15 days is one of the major constraints in rice production, mainly in rain-fed lowland areas. Floating rice cultivation is a technique of nurturing rice with the media of rafts to adapt with flood. This is feasible to be further developed on the fields flooded in years because this can maintain rice productivity during the rainy season, with potential production of 5–6 tons/ha.

Rapid intensification of Vietnamese rice production has had a positive effect on the nation's food production and economy. However, the sustainability of intensive rice production is increasingly being questioned within Vietnam, particularly in major agricultural provinces, such as An Giang. The construction of high dykes within this province, which allow for complete regulation of water onto rice fields, has enabled farmers to grow up to three rice crops per year. However, the profitability of producing three crops is rapidly decreasing as farmers increase their use of chemical fertilizer inputs and pesticides. Increased fertilizer inputs are partly used to replace natural, flood-borne, nutrient-rich sediment inputs that have been inhibited by the dykes, but farmers believe that despite this, soil health within the dyke system is degrading. However, the effects of dykes on soil properties have not been tested. Therefore, a sampling campaign was conducted to assess differences in soil properties caused by the construction of dykes. The results show that, under present fertilization practices, although dykes may inhibit flood-borne sediments, this does not lead to a systematic reduction in nutrients that typically limit rice growth within areas producing three crops per year. Concentrations of total nitrogen, available phosphorous, and both total and available potassium, and pH were higher in the surface layer of soils of three crop areas when compared to two crop areas. This suggests that yield declines may be caused by other factors related to the construction of dykes and the use of chemical inputs, and that care should be taken when attempting to maintain crop yields. Attempting to compensate for yield declines by increasing fertilizer inputs may ultimately have negative effects on the yield.

Characterized by low topography, the Mekong Delta (MD) is subjected to flooding caused by high river discharge, tidal backwater effects, and storm surges. Floods in the Mekong Delta are annual events, mainly triggered by the Asian monsoons, but also by tropical cyclones (typhoons). On the positive side, floods bring various benefits to the MD with an estimated annual value of USD 8–10 billion (MRC, 2012). These benefits include provision of sediment to counter delta subsidence, increase in wild fish catch and enhancement of soil fertility through deposited sediment (Manh et al., 2014). On the other hand, extreme floods can result in extensive damages as recorded during the floods in 2011 and 2000. For example, the 2000 flood, considered as a 20-year flood (Le et al., 2007), resulted in over 450 fatalities and economic losses of US$ 250 million (MRC, 2012). Recent studies suggest that the frequency of such extreme events is likely to increase (Delgado et al., 2010; Hirabayashi et al., 2013). For instance, the 100-year flood in the Mekong basin in the 20th century is projected to occur every 10–20 years in the 21st century due to impacts of climate change (Hirabayashi et al., 2013). Therefore, assessing hazards and risks induced by extreme floods is a crucial task for developing flood management strategies and climate change adaptation measures.

Flooding is an imminent natural hazard threatening most river deltas, e.g., the Mekong Delta. Appropriate flood management is thus required for sustainable development of the often densely populated regions. Recently, the traditional event-based hazard control shifted towards a risk management approach in many regions, driven by intensive research leading to new legal regulation on flood management. However, a large-scale flood risk assessment does not exist for the Mekong Delta. Particularly, flood risk to paddy rice cultivation, the most important economic activity in the delta, has not been performed yet. Therefore, the present study was developed to provide the very first insight into delta-scale flood damages and risks to rice cultivation. The flood hazard was quantified by probabilistic flood hazard maps of the whole delta using a bivariate extreme value statistics, synthetic flood hydrographs, and large-scale hydraulic model. The flood risk to paddy rice was then quantified considering cropping calendars, rice phenology, and harvest times based on a time series of enhanced vegetation index (EVI) derived from MODIS satellite data, and a published rice flood damage function. The proposed concept provided flood risk maps to paddy rice for the Mekong Delta in terms of expected annual damage. The presented concept can be used as a blueprint for regions facing similar problems due to its generic approach. Furthermore, the changes in flood risk to paddy rice caused by changes in land use currently under discussion in the Mekong Delta were estimated. Two land-use scenarios either intensifying or reducing rice cropping were considered, and the changes in risk were presented in spatially explicit flood-risk maps. The basic risk maps could serve as guidance for the authorities to develop spatially explicit flood management and mitigation plans for the delta. The land-use change risk maps could further be used for adaptive risk management plans and as a basis for a cost–benefit of the discussed land-use change scenarios. Additionally, the damage and risk maps may support the recently initiated agricultural insurance programme in Vietnam (Fig. 2.3).

Climate change is unequivocal. Farmers are increasingly vulnerable to floods and drought. In this article, the negative impact of climate hazards on rice cultivation in the Tonle Sap and Mekong river influenced by climatic variability between 1994 and 2018 are analyzed. A cohort of 536 households from four Cambodian districts participated in household surveys designed to consider how various vulnerability factors interacted across this time series. It was found that: (i) the major climate hazards affecting rice production between 1994 and 2018 were frequent and extreme flood and drought events caused by rainfall variability; (ii) in 2018, extreme flood and drought occurred in the same rice cultivation cycle. The impact caused by each hazard across each region were similar; (iii) an empirical model was used to demonstrate that drought events tend to limit access to irrigation, impact rice production, and result in an increased prevalence of water-borne diseases. Flood events cause reduced rice production, damage to housing, and impede children from

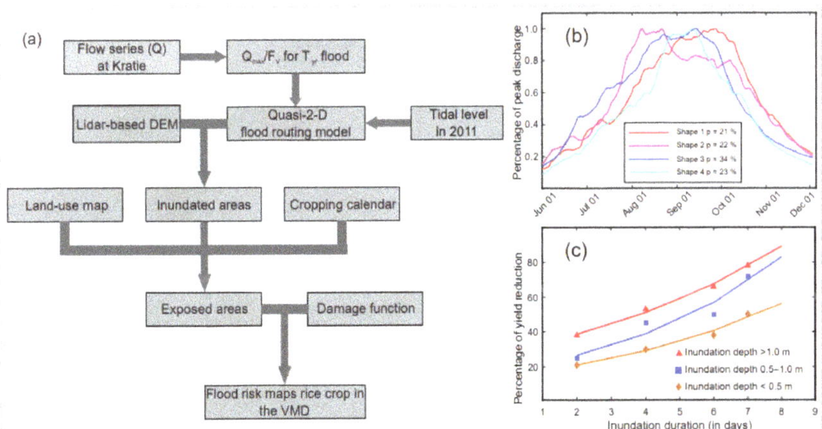

Fig. 2.3: (a) Procedure for estimating flood risk to rice production in the Vietnamese Mekong Delta; **(b)** the four normalized discharge hydrographs at Kratie, together with their probability of occurrence (Dung et al., 2015) used for the derivation of synthetic flood events as the upper boundary of the hydraulic model;. **(c)** stage-damage curve for paddy rice (Dutta et al., 2003).

accessing education. The impact of drought events on rice production was found to be more severe than flood events; however, each climatic hazard caused physical, economic, social, and environmental vulnerabilities. It is recommended that sufficient human and financial resources are distributed to local authorities to implement adaptation measures that prepare rice farmers for flood and drought events and promote equitable access to water resources (Sok et al., 2021) (Fig. 2.4).

In developing countries in general and in Vietnam in particular, flood-induced economic loss of agriculture is a serious concern since the livelihood of large populations depends on agricultural production. The objective of this study was to examine if climate change would exacerbate flood damage to agricultural production with a case study of rice production in Huong Son district of Ha Tinh Province, north-central Vietnam. The study applied a modeling approach for the prediction. Extreme precipitation and its return periods were calculated by the Generalized Extreme Value Distribution method using historical daily observations and output of the MRI-CGCM3 climate model. The projected extreme precipitation data was then employed as an input of the Mike flood model for flood modeling. Finally, an integrated approach employing flood depth and duration and

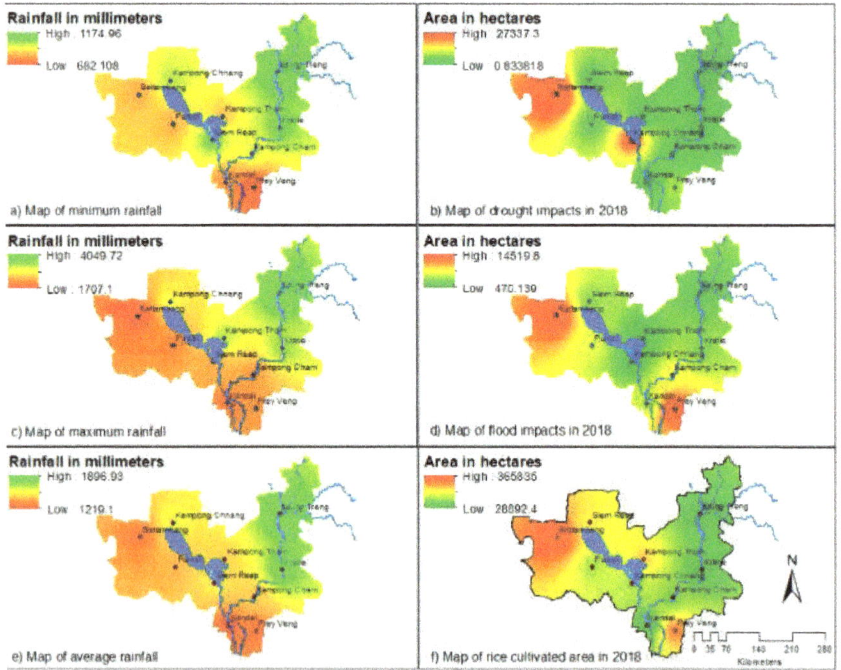

Fig. 2.4: Vulnerability to climate change on the cultivated areas of rice paddies in the Mekong river region and the Tonle Sap region.

crop calendar was used for the prediction of potential economic loss of rice production. Results of the study show that in comparison with the baseline period, an increase of 49.14 per cent in the intensity of extreme precipitation was expected, while the frequency would increase five times by 2050s. As a result, the seriousness of floods would increase under climate change impacts as they would become more intensified, deeper and longer, and consequently the economic loss of rice production would increase significantly. While the level of peak flow was projected to rise nearly 1 m, leading the area of rice inundated to increase by 12.61 per cent, the value of damage would rise by over 21 per cent by 2050 s compared to the baseline period. The findings of the present study are useful for long-term agricultural and infrastructural planning in order to tackle potential flooding threats to agricultural production under climate change impacts (Fig. 2.5 and Table 2.3).

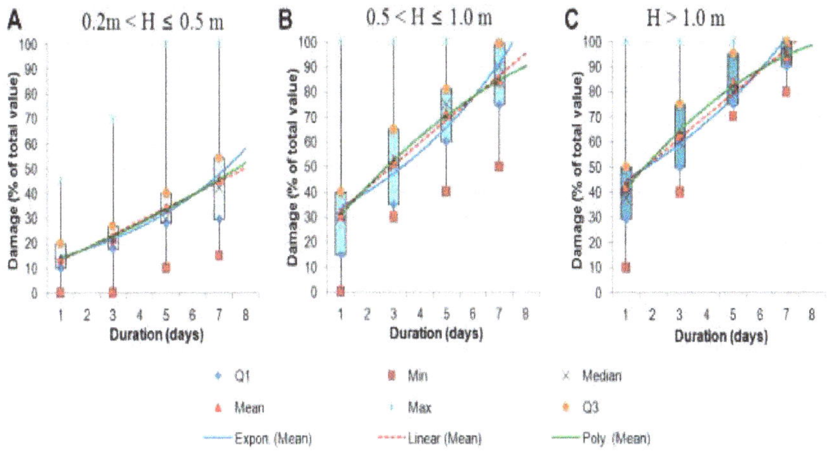

Fig. 2.5: Loss curves for rice production.

Table 2.3: Loss functions for rice production

Depth (m) Function	Exponential Function	Polynomial Function	Linear
Level 1	$y = 12.14e0.196x$ $R^2 = 0.989$	$y = 0.188x2 + 3.804x + 10.01$ $R^2 = 0.990$	$y = 5.31x + 7.942$ $R^2 = 0.995$
Level 2	$y = 28.80e0.166x$ $R^2 = 0.948$	$y = -0.484x^2 + 12.88x + 18.48$ $R^2 = 0.992$	$y = 9.001x + 23.821$ $R^2 = 0.996$
Level 3	$y = 39.42e0.134x$ $R^2 = 0.945$	$y = -0.668x^2 + 14.18x + 28.05$ $R^2 = 0.990$	$y = 8.833x + 35.413$ $R^2 = 0.992$

References

Blöschl, G., Kiss, A., Viglione, A., Barriendos, M., Böhm, O., Brázdil, R. et al. (2020). Current European flood-rich period exceptional compared with past 500 years. *Nature*, 583(7817): 560–566. Doi: 10.1038/s41586-020-2478-3. Epub 2020 Jul 22. PMID: 32699397.

Delgado, J.M., Apel, H., and Merz, B. (2010). Flood trends and variability in the Mekong river. *Hydrol. Earth Syst. Sci.*, 14: 407–418; https://doi.org/10.5194/hess-14-407-2010.

Dung, N.V., Merz, B., Bárdossy, A., and Apel, H. (2015). Handling uncertainty in bivariate quantile estimation – An application to flood hazard analysis in the Mekong Delta. *J. Hydrol.*, 527: 704–717.

Dutta, D., Herath, S., and Musiake, K. (2003). A mathematical model for flood loss estimation. *J. Hydrol.*, 277: 24–49.

Hirabayashi, Y., Mahendran, R., Koirala, S., Konoshima, L., Yamazaki, D., Watanabe, S. et al. (2013). Global flood risk under climate change. *Nat. Clim. Change*, 3: 816; https://doi.org/10.1038/nclimate1911.

IPCC *Climate Change (2007): Synthesis Report*; Intergovernmental Panel on Climate Change: Geneva, Switzerland, 2007.

Le, T.V.H., Nguyen, H.N., Wolanski, E., Tran, T.C., and Haruyama, S. (2007). The combined impact on the flooding in Vietnam's Mekong river delta of local man-made structures, sea level rise, and dams upstream in the river catchment. *Estuar. Coast. Shelf Sci.*, 71: 110–116. https://doi.org/10.1016/j.ecss.2006.08.021.

Manh, N.V., Dung, N.V., Hung, N.N., Merz, B., and Apel, H. (2014). Large-scale suspended sediment transport and sediment deposition in the Mekong Delta. *Hydrol. Earth Syst. Sci.*, 18: 3033–3053; https://doi.org/10.5194/hess-18-3033-2014.

MRC. (2012). *The Impact and Management of Floods and Droughts in the Lower Mekong Basin and the Implications of Possible Climate Change*, Mekong River Commission, 129 pp.

Reboita, M.S., Gan, M.A., Rocha, R.P., and Ambrizzi, T. (2010). *Regimes de pre-cipitação na América do Sul: uma revisão bibliográfica. Rev. Bra-sil Meteorol.*, 25: 185-204.

Sok, S., Chhinh, N., Hor, S., and Nguonphan, P. (2021). Climate change impacts on rice cultivation: A comparative study of the tonle sap and mekong river. *Sustainability*, 13: 8979.

3

Diversity of Submergence

Submergence affects crops in low-lying areas. Submergence can be defined on the basis of duration, depth, and frequency. Flooding causes submergence and damage to rice crops. Two types of environment cause submergence: flash flooding and deepwater. Flash flood submergence is defined by water levels rising rapidly and plants remaining submerged for one to two weeks. Deepwater submergence is defined by water depths greater than 100 cm persisting for months.

Submergence is a major problem in many rice-growing countries of Southeast and South Asia, particularly along the coast of Bangladesh, India, and Vietnam. Flooding has two distinct classes and mechanisms: flash flooding of short duration, in which cessation of elongation and energy-saving allows growth to restart when floodwaters retreat (e.g., QTL *SUB1*); and deep-water flooding to several metres for several months, in which plant elongation is required for foliage to remain above the water to allow respiration and photosynthesis (e.g., QTL *Snorkel*) (Sripongpangkul et al., 2000; Das et al., 2005; Hattori et al., 2011).

Transient complete submergence of less than two weeks caused by flash floods with heavy rainfall is common and affects over 20 Mha of rice (Mackill et al., 2012). Submergence is common during germination in direct-seeded rice fields, but also causes severe problems after seedling emergence in both transplanted and direct-seeded crops when the plants are still short.

Rice plants may experience submergence stress in various phases of plant development due to unpredictable flash floods occurring during planting season. Plants' recovery phase after submergence is an important phase that determines plant survival. The character of plant growth and development during this recovery phase can be an important information for determining the tolerance of plants to submerged stress. Therefore, a study was conducted to evaluate the recovery phase of post-submerged rice plants indicated by the character of plant growth and development.

Split plot design was utilized with rice cultivars as the main plot and submergence time subplot. Two cultivars used were IR 42 (non-tolerant cultivar) and Inpari 30 (tolerant cultivar), while submergence time consisted of three phases, namely: (1) generative initial phase = 42 days after transplanting (DAT); (2) active tillering phase = 28 DAT, and (3) early vegetative phase = 7 DAT. Plant samplings were conducted six times: right before submergence (t0), right after submergence (t1), four days after submergence (t2), seven days after submergence (t3), 10 days after submergence (t4) and 14 days after submergence (t5). Results showed that both cultivar and submergence time showed a non-significant effect on plant growth during recovery phase, except for leaf greenness parameter. The recovery pattern of rice plants after experiencing submergence stress was generally not influenced by the age of the plants during stress. In addition, improvements in plant growth characters began to appear after 10 days of recovery.

3.1 Submergence Ecology

Results suggest that complete absence of oxygen in floodwater will have a major adverse effect on rice plants during submergence, and that the rice plant's ability to adapt to anoxic conditions is an important component of submergence tolerance. These findings are discussed in relation to metabolic adaptations to alleviate the adverse effects of anoxia during submergence.

Submergence reduces both the radiation available for photosynthesis and the oxygen level. The damage to the submerged plants is exacerbated with increased submergence depth, higher temperature, and increased turbidity because these conditions reduce photosynthesis and increase respiration requirement. When susceptible rice plants are submerged, the level of ethylene increases, plants elongate, and chlorophyll gets degraded. This consumes energy and carbon and hastens plant mortality under transient submergence, particularly when the photosynthetic rate is reduced severely (Mackill et al., 2012).

Some landraces can withstand submergence for up to two weeks, while susceptible varieties commonly die in complete submergence after one week. Owing to the frequency of the problem in these areas, improved varieties with high-yield potential are not grown because of their susceptibility to submergence. From one of the tolerant landraces in India, FR13A was selected, and the QTL for submergence tolerance (*SUB1*) was identified on chromosome 9 (Mackill et al., 2012). Expression of *SUB1* is induced by ethylene, which inhibits plant elongation and hence saves energy within the plants, resulting in survival for about two weeks. *SUB1* is effective from four days after germination to two weeks before flowering. Marker-assisted backcrossing was used to introgress the gene into several mega varieties in South and Southeast Asia. These *SUB1* varieties behave like the mega varieties when there is no flood in terms of grain yield and grain quality but survive submergence much longer. Comparisons of the mega varieties with or without *SUB1* show the yield advantages of *SUB1* varieties from 1 to more than 3 t ha^{-1} (Ismail et al., 2013). These *SUB1* introgression varieties were adopted very rapidly, covering areas exceeding 1 Mha by 2012.

In addition to flash flooding, long-term partial stagnant flooding is common in low-lying areas. There are some varieties which are tolerant to long-term partial stagnant flooding, but this is not related to *SUB1* (Singh et al., 2011). Deep-water and floating rice requires an entirely different strategy because prolonged flood depth and duration precludes the tolerance strategy. Here, stems must elongate with the rising floodwater so that oxygen can continue to be transported to the roots. Submergence causes changes in gene expression, which coordinate morphological and metabolic adaptations to the stress (Bailey-Serres and Voesenek, 2008). In rice-growing areas, crops frequently face floods which may be a flash flood and complete submergence. Transcription factors, such as group VII ethylene response factors (ERF-VIIs) regulate the expression of a varied range of genes involved in adaptive responses to excess moisture or low oxygen (Voesenek and Bailey-Serres, 2015). In response to submergence, deepwater paddy enhances cell division and elongation growth by ethylene-mediated pathway through activation of ACC synthase gene. Various genes, like SNORKEL1 and SNORKEL2 (ERF VII) were also reported in deepwater rice (Hattori et al., 2009) that encode two ethylene-responsive factors (ERF) DNA-binding proteins. In lowland rain-fed paddy, escape and quiescence strategies are followed to get rid of hypoxia conditions due to flooding. In rice, an ERF-VII TF gene (SUB1A) is known as the main regulator of submergence tolerance, permitting plants to withstand complete submergence for 14–16 days (Fukao et al., 2006; Xu et al., 2006). It is demonstrated that SUB1A enhances inhibition of gibberellic acid (GA) signalled by the upregulation of rice DELLA proteins, like SLR1 (Slender Rice 1) and SLRL1 (SLR1-like1), known as negative regulators of GA (Fukao and Bailey-Serres, 2008). Consequently, GA signalling is inhibited, causing limited elongation of growth.

Low oxygen also regulates the function of various K^+ channels in mammals (Wang et al., 2017). Under waterlogging, K^+ concentration in the soil considerably drops and the uptake of K^+ by plants is restricted due to reduced hydraulic conductivity. Further, this suggests that low oxygen and flooding can increase internal K^+ concentration in specific tissues and cell types due to altered ion channel activities. A recent study noted that under combined oxygen-deprived and K^+ sufficient conditions, there was the accumulation of AtRAP2.12 protein (an *Arabidopsis* ERF-VII) improved by a Raf-like mitogen-activated protein kinase kinase kinase (MAPKKK) and HCR1 (Shahzad et al., 2016).

Although rice plants require large amounts of water during growth, flooding stress results in severe loss of the crop as a consequence of anaerobic environment. The crop loss would be more severe on account of unpredictable changes in climatic conditions causing floods (Vergara et al., 2014). The extensive rice-growing areas in South and Southeast Asia, especially India, Bangladesh, Thailand, Vietnam, Myanmar and Indonesia are exposed to flash flooding during the monsoon season (Sarkar et al., 2006). Nearly 22 million hectares of rain-fed lowland areas of South and Southeast Asia get affected due to flooding, out of which about 6.2 million hectares of rice lands are in India (Azarin et al., 2017; Dar et al., 2017). Out of 22 million hectares, 15 million hectares of rain-fed lowland are affected only due to short-term flash flooding (Singh et al., 2017), and economic loss is estimated to be 1 billion US dollar (Mackill et al., 2012). The flood-prone ecosystem comprises about 7% of global rice area and produces 4% of world rice (Yang et al., 2017).

3.2 Types of Flooding Stress

3.2.1 Flooding during Germination

Soil waterlogging normally takes place in case of rainfall after sowing of seeds and particularly where the lands are not properly levelled. Germination during flooding is commonly referred as 'Anaerobic Germination' (AG), and the seeds have the potential to germinate even under the conditions where no air or oxygen is available. However, slow seed germination, uneven and delayed seedling establishment and high weed infestation are some of the constraints that restrict rice adaptation of direct seeding in flood-prone areas (Ismail et al., 2009, 2012). Almost all varieties of rice are keenly susceptible to flooding and get easily affected or damaged at the time of germination (Yamauchi et al., 1993; Ismail et al., 2009; Angaji et al., 2010; Miro and Ismail, 2013; Lee et al., 2014; Singh et al., 2017). AG is a complex trait governed by various genes which are linked to many important processes, like starch breakdown, glycolysis, fermentation as well as other biochemical and metabolic processes (Ismail et al., 2009, 2012). Hence, AG is thus necessary for homogeneous germination and better crop establishment under flooding conditions (Ismail et al., 2009; Magneschi and Perata, 2009; Septiningsih et al., 2013).

3.2.2 Flash Flooding

Flash flooding can occur due to heavy rains or overflowing of rivers and streams for a short duration of one or two weeks, resulting in complete submergence of plants (Vergara et al., 2014). Lands in low-lying areas and near streams and rivers are generally affected by such flash flooding. Depending on climatic conditions, flash flooding can occur repeatedly in a growing season, thus damaging the rice cultivation. The flooding depth is not very deep in the flash flooding condition. Flooding tolerance was defined as 'the ability of some rice varieties to survive 10–14 days of complete flooding and renew its growth when the flood water subsides' by Catling (1992).

3.2.3 Stagnant Flooding

In stagnant flooding, water depth ranges from 25 to 50 cm, and some parts of the shoot is normally seen over the flood-waters. Rice production in such cases depends upon the extent of submergence and may vary from 0.5 to 1.5 t/hm^2 (Kuanar et al., 2017). In this condition, the extended water stagnation varies from a few weeks to several months. This is commonly seen in areas where there is no proper drainage system, particularly near the water channels leading to rivers during the monsoons due to overflow of excessive rain-water. In West Bengal of India, 577 genotypes were studied under stagnant flooding conditions, where genotypic differences between survival and grain yield were observed, but the flooding tolerant cultivars (FR13A and FR43B) were not found to perform well under the aforesaid condition. This suggests that flash flooding-tolerant genotypes may not be the right solution for stagnant flooding (Vergara et al., 2014).

3.2.4 Deep-water Flooding

In deep-water flooding, water stagnates for a longer time. At times, the water level may rise up to 4 m height (Singh et al., 2017). The flooding stress, depending upon the topography and climatic conditions, may continue for months together. Adapted rice plants of this type accelerate their growth in commensurate with the increase of flood water level, so as to avoid complete submergence. In deep-water flooding, a portion of plants successfully remains above the water level. The deep-water rice escapes complete flooding by fast internodal elongation growth, which consumes high amounts of carbohydrate. Growth is achieved by about 25 cm per day as floodwater level increases (Singh et al., 2017) and can reach a height of 5 m with its panicle and top leaves under these circumstances (Jackson and Ram, 2003). Most of the deep-water rice-growing areas in India are normally found in several parts of Assam, Bihar, Orissa, West Bengal and Uttar Pradesh (Bin Rahman and Zhang, 2016).

Flooding causes many complex abiotic stresses (Jackson and Ram, 2003; Sarkar et al., 2006; Bailey-Serres et al., 2010), and the amount of harm that may be caused to the inundated plants varies, depending on floodwater characteristics, like temperature, turbidity, water depth, oxygen, carbon dioxide concentration and light intensity (Das et al., 2009). These affect important plant processes, like chlorophyll retention, photosynthesis under water, accumulation of carbohydrate, elongation and survival (Das et al., 2009).

The genetic basis of flooding tolerance remained ambiguous until mid-1990 (Bailey-Serres et al., 2010). Genetic mechanism of rice for tolerance to flooding/submergence is shown in Fig. 3.1. 'Quiescence' and 'elongation' are two major and opposite strategies to encounter the effects of flooding during flash and deep-water flooding, respectively (Fig. 3.1). The 'escape' type of adaptive response is mostly seen in deep-water lowland rice, where vigorous shoot elongation prevents the plant from complete flooding as flood-water level gradually increases. Two ethylene responsive factor genes, SNORKEL1 (SK1) and SNORKEL2 (SK2), cloned from a deep-water rice variety C9285 are involved in the elongation of shoot in the floating rice (Vergara et al., 2014; Tamang and Fukao, 2015). Both the genes function in contradiction with Submergence1A-1 (Sub1A-1). The Sub1A-1 empowers the plant to endure flash flooding by

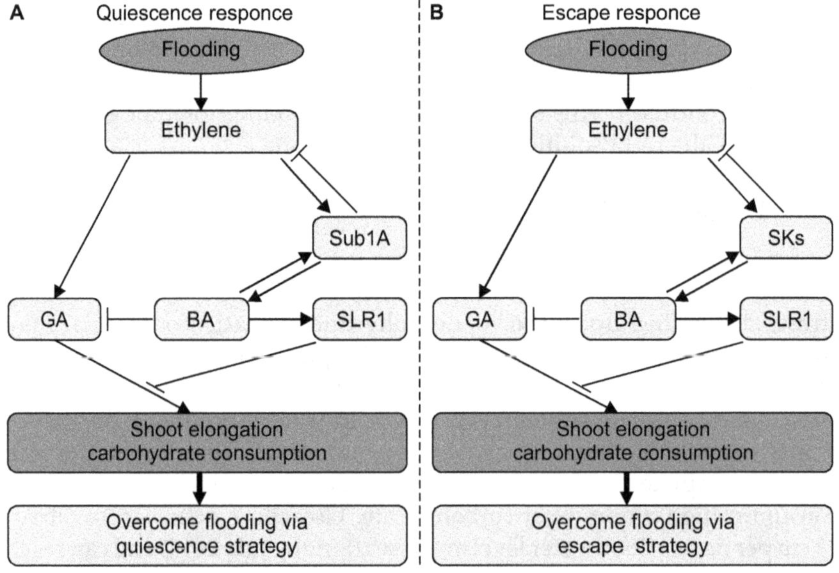

Fig. 3.1: Genetic mechanism of rice for tolerant to flooding/submergence. A, Sub1A mediated response of submergence tolerance in rice. B, SNORKEL (SK) mediated escape response for tolerance to deep-water flooding in rice (Tamang and Fukao, 2015).

	Submergence tolerant (*Sub1*)	Lowland submergence intolerant	Deepwater rice
Waterlogged paddy depth (cm)	5 to < 50	5 to < 50	> 50 to 400
Subspecies	*aus*, *indica*	*japonica*, *indica*	*indica*
Underwater response strategy	quiescence	escape	deepwater escape
Ethylene-triggered GA-mediated elongation	suppressed	enhanced in shoot, leaf	enhanced in internodes
ERF controlling response strategy	*SUB1A*		*SK1, SK2*

limited elongation until flooding-water recedes (Xu et al., 2006; Bailey-Serres et al., 2010). The tolerant rice genotype adapted to flash flooding condition is characterised by slow elongation growth and thus conserves accumulated respirable biomass to resume growth after de-submergence (Bailey-Serres et al., 2012; Azarin et al., 2017). Allelic surveys showed that SNORKEL (SK) genes are present only in deep-water rice varieties which exhibit rapid internode elongation in response to submergence (Tamang and Fukao, 2015). SKs promote the accumulation of bioactive gibberellic acid in submerged internodes (Ayano et al., 2014). During flooding, ethylene accumulates, triggering SK1/2 gene expression in C9285 (Minami et al., 2018). It has been observed that ethylene increases biosynthesis of gibberellic acid under flooding, triggering internode elongation in deep-water rice (Fukao and Bailey-Serres, 2008). Kuroha et al. (2018) identified rare allele of gibberellin biosynthesis, a gene SD1 (SEMIDWARF1), which provides adaptation to deep water. The SD1 protein directs increased gibberellin synthesis, largely GA_4, thus resulting in internode elongation.

3.3 Search for Suitable Varieties

An estimated 10 million hectares in Bangladesh and India alone are marginalized by the threat of flooding during each monsoon season (Huke and Huke, 1997). As a coping strategy, farmers have traditionally cultivated chronically flood-prone lowlands with landraces that can endure 10 days or more of complete submergence and resume growth upon de-submergence (Catling, 1992). However, these submergence tolerant landraces produce less than 2 t of grain ha^{-1}, paling in comparison to the 6–8-t of grain ha^{-1} yields of advanced semi-dwarf varieties. Unfortunately, the popular 'mega-varieties' grown in large areas of Asia are sensitive to complete submergence and usually die within seven days of complete inundation.

The major benefit of using the MABC approach is the resultant Sub1 varieties which retain all the desirable features of the recurrent parent, especially the yield and quality characteristics. One explanation for the lack of adoption of previously developed Sub1 varieties, such as IR49830-7- 1-2-2 (Mackill, 2006), was that genomic regions from the non-recurrent parent remained. Varieties with the Sub1 region from FR13A have the same yield, and other agronomic and grain quality characteristics as the original varieties when grown under shallow paddy conditions in the field; however, when subject to complete submergence for seven to 15 days, these varieties showed considerable yield advantages (Singh et al., 2009; Sarkar et al., 2009). With the Sub1 mega-varieties, dissemination and adoption is more straightforward because the main aim is the replacement of the original mega variety with an improved submergence-tolerant version. The adoption of the Sub1 mega-varieties in non-flood-prone areas is likewise not a problem. Importantly, these introgression lines can also replace some of the low-yielding traditional landraces currently being used by farmers in submergence-prone areas, augmenting yields in typically marginalized fields.

Submergence interrupts all oxygen-dependent mechanisms inside the plant cells by shifting aerobic respiration to anaerobic respiration (Ayi et al., 2016). The level of respiration is directly related to temperature, e.g., increase in temperature causes increase in respiration resulting in reduced oxygen inside the cells (Colmer et al., 2011). Submerged rice plant faces energy crisis due to the low level of O_2 (anoxic condition) which limits the respiration, ultimately causing plant death (Crawford and Braendle, 1996). In submerged plants, reduced photosynthesis and enhanced consumption of photo-assimilates in anaerobic respiration (fermentation) also lower the carbohydrates reserves, and such imbalance and over-consumption of sugars during submergence may lead to complete failure of plant growth or even plant death (Bailey-Serres and Voesenek, 2008; Colmer and Voesenek, 2009).

Flooding or submergence is one of the major environmental stressors affecting many man-made and natural ecosystems worldwide. Increase in the frequency and duration of heavy rainfall due to climate change has negatively affected plant growth and development, which eventually causes the death of plants if the condition persists for days. Most crops, especially rice, being a semi-aquatic plant, are greatly affected by flooding, leading to yield losses each year. Genetic variability in the plant response to flooding includes the quiescence scheme, which allows underwater endurance of a prolonged period, escape strategy through stem elongation, and alterations in plant architecture and metabolism. Investigating the mechanism for flooding survival in wild species and modern rice has yielded significant insight into developmental, physiological, and molecular strategies for submergence and waterlogging survival. Significant progress in the breeding of submergence-tolerant rice varieties has been made during the last decade following the successful identification and mapping of a quantitative trait locus for submergence tolerance, designated as SUBMERGENCE 1 (Sub1) from the FR13A landrace. Using marker-assisted backcrossing, the *SUB1 QTL* (quantitative trait locus) has been incorporated into many elite varieties within a short span of time and with high precision as compared with conventional breeding methods. Despite the advancement in submergence tolerance, for future studies, there is a need for practical approaches to explore genome-wide association studies (GWA) and QTL in combination with specific tolerance traits, such as drought, salinity, disease and insect resistance.

The major objective of the international community is to increase the current agricultural production to catch up with the anticipated growth in population. Based on this background, it is essential to adopt germplasm and develop better abiotic resistant cultivars that can withstand flooding. Achieving this objective was reflected in the development of submergence-tolerant rice varieties that survive floods, and were subsequently deployed to farmers' fields in flood-prone rice-cultivating areas. This was manifested through the International Rice Research Institute's (IRRI) initiation of a robust submergence tolerance breeding programme of modern rice varieties that were developed through marker-assisted selection from local landraces. Therefore, understanding the molecular, physiological, morphological, and developmental mechanisms that trigger flooding tolerance would enable the translation of survival strategies of rice flooding to other crops to stabilize crop yields.

Rice is widely grown in varied environmental conditions, ranging from sea-coasts to high altitudes. The major rice-growing areas are greatly affected by flooding caused by river discharge, excessive rain-water accumulation, and tidal movements. Globally, one-third of rice-cultivated areas are deep-water and rain-fed lowland ecosystems, which account for about 50 million hectares (Bailey-Serres and Voesenek, 2008). These areas are prone to frequent flooding, which could be attributed to poor drainage systems of excessive rain-water during the rainy season. Rice has anaerobic tolerance; however, extreme flooding, either complete or partial submergence, may result in various environmental stressors. Hence, rice cultivars respond differently to the variations in the flood-water regime. Therefore, it can be concluded that a flood-tolerant cultivar at one location cannot be transferred to another location (Dar et al., 2017). There are three categories of flooding that occur at different crop growth stages and varied durations, which are typical of rice-growing agro-ecosystems.

3.4 Low Gas Diffusion

Waterlogging is defined as a condition of the soil in which excess water limits gas diffusion (oxygen diffusivity in water is approximately 10,000 times slower than in air, and the flux of O_2 into soil is approximately 320,000 times less when the soil pores are filled with water than when they are filled with gas) (Armstrong and Drew, 2002; Colmer and Flowers, 2008). The principal cause of damage to plants grown in waterlogged soil is inadequate supply of oxygen to the submerged tissues as a result of slow diffusion of gases in water and rapid consumption of O_2 by soil microorganisms. Oxygen deficiency in waterlogged soil occurs within a few hours under certain conditions. In addition to the O_2 deficiency, production of toxic substances, such as Fe^{2+}, Mn^{2+}, and H_2S by reduction of redox potential causes severe damage to plants under waterlogged conditions (Drew and Lynch, 1980; Setter et al., 2009). Thus, growth and development of most plants, except for rice (*Oryza sativa* L.) and other wetland species are impeded under waterlogged conditions.

3.5 Strategies for Adaption

To a great extent, light intensity, turbidity, temperature, pH, gas diffusion, aerenchyma, and leaf morphology are the most significant factors that influence plant growth and survival. During submergence, light intensity reaching the leaves gets diminished by suspended phytoplankton or silt and dissolved organic matter in the water. Similarly, during flash floods, the water is mostly turbid and only a meagre amount of solar radiation can reach the plant canopy, thereby limiting the photosynthesis rate. Under submergence, light intensity is a major factor in controlling the concentration of carbon dioxide and oxygen, which have a significant impact on the physiological status of the submerged plant (Sarkar et al., 2006). Generally, plants produce oxygen through photosynthesis. However, a lack of effective oxygen transport system to non-photosynthesis organs infers that these organs dispossess oxygen if their physiological mechanism restricts the diffusion of oxygen to the outside (Loreti et al., 2013). In addition, plants respond differently to insufficient oxygen, limiting respiration (hypoxia), and a complete absence of oxygen (anoxia), which is detrimental to plant survival (Loreti et al., 2013). Flooding conditions can drastically affect plant survival. As reported by Pucciariello et al. (2014), large survival differences exist between completely submerged plants in darkness versus those submerged with some light, suggesting the importance of photosynthesis under water. Waterlogging hampers the development and function of roots and shoots due to shortage of oxygen, which restricts respiration. The consequences of photosynthesis blockage and root aerobic respiration are a result of the light intensity of water, CO_2, and O_2 (Voesenek and Bailey-Serres, 2013). Oxygen deficiency and low light intensity are the two major factors that limit the survival ability of rice plants under prolonged submergence. These are due to the failure of plant sensitivity to develop new leaves and the severe damage caused on the older leaves. Furthermore, the submerged plants are often deprived of oxygen and high light tension, which results in the formation of reactive oxygen species (ROS), such as hydrogen peroxide, hydroxyl radical, and superoxide anion that, if not moderated, can seriously damage cellular organization, leading to plant death. However, rice cultivars that are efficient in ROS detoxification after de-submergence are characterized by the ability to retain chlorophyll, sustain plant growth, regenerate new leaves, and maintain the older leaves. Hence, rice plants protect themselves against oxidative damage by using two mechanisms: (i) the presence of antioxidant enzyme systems, (ii) natural antioxidants. Low-molecular-weight naturally occurring antioxidant compounds, such as phenols, a-tocopherol, ascorbate, carotenoids, and glutathione, are found to have reduced oxidative damage in plants. However, among these compounds, only ascorbate has been studied. Hancock and Viola (2005) and Das et al. (2004) reported (Fig. 1) scale of the rice-growing environment with respect to water level: (a) Upland rain-fed; (b) lowland rain-fed (groundwater); (c) flood recession irrigated; (d) flood-prone irrigated (rice field); (e) deep-water (floating rice). Submergence during germination is the first category of floods which is often referred to as anaerobic germination (AG). This condition commonly occurs in rain-fed lowland ecosystems, which are characterized by heavy and sudden rainfall mostly associated after direct seeding cultural practices. Under this condition, rice seeds are submerged completely and suffer anoxia or hypoxia, which results in low or poor germination, poor crop establishment, and seedling death (Ismail et al., 2008). However, Angaji et al. (2010) reported rice germplasm with the ability to germinate under anoxia and hypoxia conditions. Complete submergence during the vegetative growth stage is the second category which occurs either at seedling or tillering stage. Plants are completely flooded, ranging from days to weeks before the flooding recedes. Singh et al. (2017) defined flooding tolerance in this case as the plant's ability to survive 10 to 14 days of complete submergence and continue their growth process after the flood recedes, with little damage to plant morphology. This problem is endemic in river-basin areas, and it often leads to complete yield loss due to flash flood. Xu et al. (2006) cloned the SUB1 gene from the FR13A of 16 landrace from India, which was adapted to submerge at the vegetative stage. The third category of flooding occurs in lowland rice cultivation, where the water level ranges from 20 to 50 cm, depending on whether it is a rain-fed or irrigated system. However, due to seasonal changes and rain duration, the water level might exceed 50 cm, up to

4 m, which lasts for a prolonged period of time and could partially or completely submerge the plant, depending on the growth stage. This type of flooding may remain for few months, depending on the area. Many landraces, also known as floating rice, are adapted to this type of flooding and can extend up to 25 cm per day as the flood level increases (Singh et al., 2017). Generally, stagnant flooding occurs as a result of heavy rain, which can be deep-water flooding or flash flooding. Under deep-water flooding, plants increase their stem length (internode) to escape the submergence. However, after the flood subsides, the affected plants become prone to logging and eventually this leads to death due to exhaustion of reserved nutrients within a few days. Contrarily, during flash flood, plant growth is restricted and resumes after occurrence of the flood (Winkel et al., 2013). Regardless of the temperature, duration, and depth of the flood, flood has a major impact and threatens plants with shortage of carbohydrates and cellular energy, which hinder growth and development (Fig. 3.2).

Rain-water flooding usually means clear water, which results in less crop damage as compared to turbid or silted water. Hence, comprehensive knowledge of the relationship between plant survival and flood water qualities is a prerequisite for the development of flood tolerance and sustainable flood-management practices. Five of 16 higher levels of ascorbate are in the root system under hypoxia, while it declines upon re-aeration. Therefore, ascorbate plays an important role in protecting plants against ROS damage. Quiescence strategy is observed by traits that conserve carbohydrates and energy under prolonged flooding conditions and recover normal growth and development after the flood subsides,

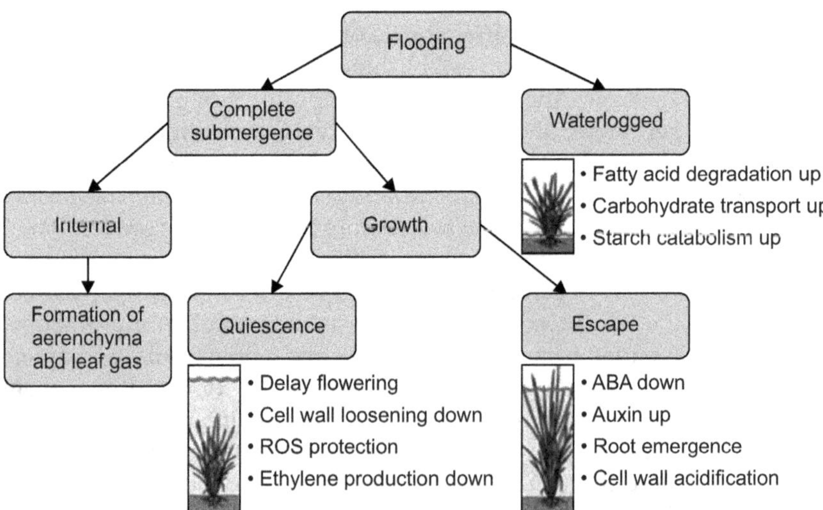

Fig. 3.2: Rice strategies for adaptation to submergence and waterlogging.

whereas escape strategy is recognized by the collection of anatomical and morphological traits that enable gaseous exchange between submerged and non-submerged organs through elongation. Kende et al. (1998) studied physiological mechanisms, and suggested that gibberellin (GA), abscisic acid (ABA), and ethylene are the major plant hormones involved in deep-water rice elongation. Schaller and Bleecker (1995) described ethylene as the only gaseous plant hormone, which controls many plant characteristics, including growth, development, and senescence. Ethylene has been extensively studied as the major factor regulating internode elongation among hormones responsible for deep-water elongation. Physiological studies have shown the relationship between deep-water elongation and ethylene response (Kende et al., 1998). During complete submergence, there is a rapid increase in intralacunar ethylene content and the application of an ethylene inhibitor suppresses the elongation of the internode (Metraux and Kende, 1983). Similarly, low oxygen sensing stimulates internode elongation (Raskin and Kende, 1984). Upon reoxygenation after a period of oxygen deprivation, ethanol, that was trapped in the tissues, is converted into acetaldehyde, which causes post-anoxic cell damage. Also, reactive oxygen species (ROS) accumulate excessively upon re-oxygenation or oxygen deprivation under light conditions (Voesenek and Bailey-Serres, 2013).

3.6 Varied Submergence

Flooding is a frequent natural calamity affecting global food supply and financial security. The intensity of rainfall events is expected to increase under future climate change scenarios, which will greatly impact rice production. Different flooding patterns can cause damage or complete yield loss in rice plants at different stages of growth. This includes (a) complete submergence due to flash flood at vegetative or pre-flowering stages, (b) stagnant flooding of medium-deep water and deepwater or floating rice, and (c) submergence at germination or anaerobic germination. Different molecular and physiological mechanisms underlie tolerance to each type of flooding. Several major QTLs have been mapped and several key genes underlying the QTLs have been cloned. Remarkable progress has been achieved through conventional and molecular breeding strategies in developing tolerant varieties to mitigate the impact of different flood events. This effort will be continued in the future by incorporating new QTLs/genes and tolerance to other abiotic and biotic stresses according to the needs of the target regions. Genetics, genomics, and other modern technologies will also be continuously explored to further our understanding of how rice plants cope with different types of flooding stress.

Rice farmers in rain-fed and irrigated areas are shifting to direct seeding from transplanted rice as it provides opportunities to reduce costs and can result in earlier harvest (Balasubramanian and Hill, 2002). There are constraints, however, that limit its large-scale adoption, the most important of them being (i) poor germination and uneven stand establishment in areas where the land is not well levelled or water is not well controlled, as in rain-fed areas, and (ii) high weed infestation (Du and Tuong, 2002). Commonly, lowland fields are not well levelled, which means that they can neither be completely drained nor flooded to an even depth to control weeds. With rainfall being unpredictable, flooding of low-lying areas can result in a severe reduction in rice establishment. This is likely to be a particular problem with monsoon-season rice crops.

The development of rice varieties that can germinate and emerge in flooded soils will therefore help reduce the hazards of early floods, to which rice is very sensitive (Ismail et al., 2009), and this also provides an efficient means for weed control through early flooding. This can be achieved through effectively exploiting the genetic variation in flooding tolerance during seed germination and early establishment. Moreover, the effectiveness of flooding for weed management will depend on the responses of various weeds associated with rice to early flooding, an area that has not been studied sufficiently.

3.7 Management Strategies

3.7.1 Seed Priming

In another set of trials (Ella et al., 2011), the influence of pre-soaking of seeds for 24 hours prior to sowing or of priming (soaking for 24 hours, followed by drying) was tested. Both treatments accelerated and improved germination and seedling survival in flooded soils, especially of the tolerant genotypes. Priming reduced lipid peroxidation and increased activity of superoxide dismutase (SOD) and catalase (CAT).

In rice, different crop management strategies can improve the resistance under submerged conditions. Seed priming is used to improve tolerance in various plant species against abiotic stresses, including heavy metal, salinity, drought, chilling, and submergence stress (Jisha et al., 2013; Paparella et al., 2015; Hussain et al., 2016a, b). Seed priming has positive effects on growth and metabolic activities, under stress as well as normal conditions (Farooq et al., 2009; Khaliq et al., 2015; Zheng et al., 2016). It ensures uniform and timely germination of rice seedlings under normal and stress conditions (Khaliq et al., 2015; Hussain et al., 2016a, b). Seed priming enhances the tolerance by increasing the activity of antioxidant defence mechanisms, carbohydrate metabolism, and seedling vigour (Ella et al., 2011) (Fig. 3.3).

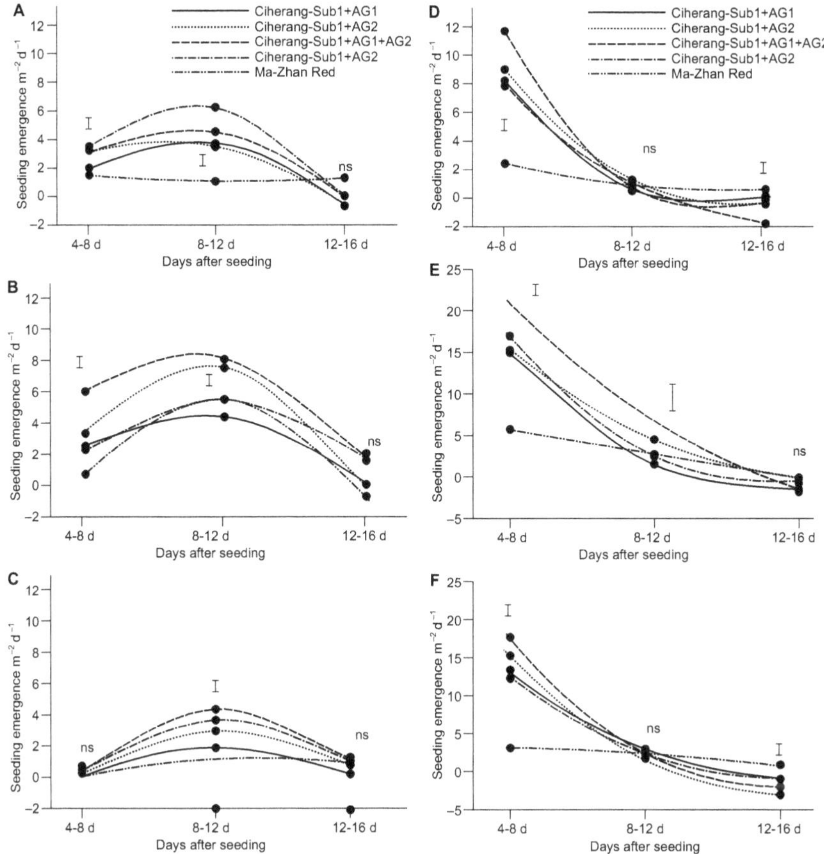

Fig. 3.3: Seedling emergence rate of AG lines under field condition with pre-treated seeds and flooding at 3–5 cm water depth, during the wet season of 2016 (**A–C**); (**A**) priming; (**B**) 24 h soaking; (**C**) dry seeds) and dry season 2017 (**D–E**); (D) priming; (**E**) 24 h soaking; (**F**) dry seeds). Vertical bars indicate LSD0.05 (n = 90).

Hussain et al. (2016a, b) found that priming with salicylic acid (SA) and selenium (Se) improved the seed germination and other morphological traits under submergence stress. Efficient use of nutrients and other inputs will enhance the productivity of transgenic rice cultivars, e.g., Khao Hlan On, Chiknal, Tilakachari, and Sirambe Putih in submerged areas of Myanmar, Bangladesh, India, and Indonesia, respectively (Vergara et al., 2014). Exogenous potassium (K) application could improve plant growth, chlorophyll contents and photosynthetic capacities as well as nutrients uptake in plants under submerged conditions (Ashraf et al. 2011). Furthermore, application of K fertilizer alone or in combination with phosphorus could maintain internal gas diffusion and energy levels required for normal growth and developmental processes under submergence conditions in rice (Singh et al., 2014a, b). Numerous novel agronomic approaches can be deployed to enhance rice tolerance for submergence, like selection of suitable rice variety, higher seed rate, seed priming, optimum sowing depth, seedling age, wider spacing, pre- and post-nutrient/amendments application and application of certain growth regulators (Munda et al., 2016; Bishnoi et al., 2017). Planting depth of more than 1 cm and beyond led to anoxic condition due to sharp decrease in redox potential whereas too much shallow sowing (0.5 cm) leads to floating of developing young seedlings above the water surface and ultimately leading to poor crop establishment (Vartapetian and Jackson, 1997; Magneschi and Perata, 2009; Miro and Ismail, 2013; Chamara et al., 2018). Wider spacing ensures higher underwater radiation, which leads to higher photosynthesis and less senescence and thus it imparts higher post-submergence recovery as compared to closer spacing (Bhaduri et al., 2020). The change in nutrient dynamics (Kumar et al., 2019) under submerged condition makes it compulsory to alter the nutrient schedule so that the submerged plant can recover from the stress quickly and optimum yield can be ensured. Innovative nutrient management practices, like application of N (post-submergence) and basal P-enhanced NSC content, survival percentage and reduced shoot elongation, which could increase plant survival and productivity under submergence (Gautam et al., 2014; Htwe et al., 2019; Mamun et al., 2017).

3.8 Molecular Approach to Develop Submergence-tolerant Cultivars

Traditional breeding approaches that comprise bulk and pedigree selection based on morphological markers have been successfully used in rice breeding for submergence-stress tolerance. Transgenesis of SUB1 gene enhances the tolerance against submergence in those rice cultivars which are already good in agronomic traits (Siangliw et al., 2003; Xu et al., 2006; Neeraja et al., 2007; Septiningsih et al., 2009, 2013; Singh et al., 2010; Thomson et al., 2010; Manzanilla et al., 2011; Mackill et al., 2012; Collard et al., 2013). Introduction of SUB1 gene amplifies the yield by two to five times following complete submergence for 12–17 days (Iftekharuddaula et al., 2011; Dar et al., 2013). Rice cultivars containing SUB1 gene complete their life cycle earlier and give more yield as compared to the non-SUB1 cultivars facing submerged conditions (Singh et al., 2009; Manzanilla et al., 2011). In future, molecular breeding approaches are required for developing superior submergence-tolerant varieties.

Recently, Fleck et al. (2011) showed that Si nutrition increased suberization and lignification of rice roots, which was accompanied by silicic acid-triggered transcription of genes associated with suberin and

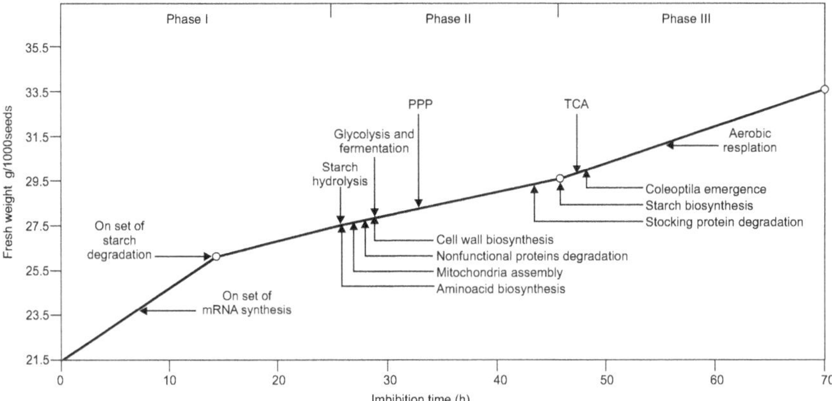

Fig. 3.4: Rice seed germination with time.

lignin biosynthesis. As a consequence of suberization and lignification of the outer root cell layers, the oxidation power of the rice roots was reduced. Although it is suggested by Fleck et al. (2011) that altered levels of silicic acid play a role in promoting the biosynthesis of suberin and lignin, the signal involved in inducible ROL barrier formation remains unclear.

Germination in rice takes place in three phases: (i) rapid water uptake phase, (ii) plateau of water uptake phase, (iii) and a further increase of water uptake phase or the initiation of growth phase (He and Yang, 2013; Narsai et al., 2013). Phases i and ii entail major metabolic reprogramming, initiating with mRNA biosynthesis and anaerobic degradation of starch at phase i followed by degradation of storage proteins at phase ii. At the end of phase ii, embryo germinates. TCA cycle starts operating during the fag end of Phase iii. Finally, at phase iii, coleoptiles growth takes place along with commencement of photosynthetic starch biosynthesis (He and Yang, 2013). Among these three phases, the second phase is most critical which initiates at about 12 hours post imbibition of water and culminates at about 48 hours post imbibition of water (Fig. 3.4). During this phase, the embryo prepares itself for germination and the food reserve stored inside the seed is transformed into a state that can support the germination process (He and Yang, 2013). In aerobic condition, oxygen uptake also follows similar trend as that of water uptake (Narsai et al., 2013). This absorbed oxygen is used for aerobic respiration which yields energy required for seed germination. But when the oxygen supply is restricted, the germination process falls back to anaerobic respiration as an alternative resort to fulfil energy demand (Narsai et al., 2013). So, the success of GSOD tolerance largely depends on the flexibility of reprogramming respiration process from aerobic to anaerobic mode (Table 3.1).

Submergence tolerance is an important aspect to be considered when flash flood damages rice. Tolerant genotypes can withstand submergence

Table 3.1: Changes of seedling establishment, length and dry weight due to germination stage submergence.

Genotype	Seedling est. %			Seedling Length (cm)			Seedling DW (mg plant^{-1})			Vigour Index
	C	T	T/C	C	T	T/C	C	T	T/C	
Naveen	99	25	0.25	33.5	20.0	0.60	36.0	18.0	0.50	500
AC41620A	99	86	0.87	29.3	29.9	1.02	46.5	25.1	0.54	2571
FR13A	98	27	0.28	20.0	17.0	0.85	42.5	24.0	0.56	459
Kamini	99	22	0.22	36.6	28.5	0.78	23.3	15.9	0.68	627
AC41647	97	41	0.42	38.5	34.8	0.90	50.6	32.4	0.64	1427
Ravana	98	15	0.15	41.8	39.5	0.94	60.6	13.9	0.23	593
AC39416A	96	35	0.36	41.3	35.4	0.86	53.3	26.6	0.50	1239
Rashpanjor	96	71	0.74	40.0	27.5	0.69	50.1	27.5	0.55	1953
Talmugra	96	34	0.35	30.9	33.4	1.08	67.4	31.8	0.47	1136
Pokkali	95	17	0.18	36.0	28.4	0.79	57.8	15.9	0.28	483
Paloi	96	21	0.22	38.9	31.7	0.81	62.0	37.1	0.60	666
Panikekua	99	23	0.23	38.3	33.3	0.87	53.7	25.9	0.48	766
Pantara	99	38	0.38	29.1	28.8	0.99	42.7	22.3	0.52	1094
AC40346	99	63	0.64	27.5	25.9	0.94	33.6	24.5	0.73	1632
JRS8	99	60	0.61	33.3	30.9	0.93	45.8	25.3	0.55	1854
JRS20	94	46	0.49	30.4	30.0	0.99	55.7	29.8	0.54	1380
JRS21	98	48	0.49	37.0	31.0	0.84	57.5	24.3	0.42	1488
JRS155	94	34	0.36	36.0	32.0	0.89	49.8	29.2	0.59	1088
JRS182	99	51	0.52	33.8	30.5	0.90	53.6	31.2	0.58	1556
JRS196	99	63	0.64	32.4	29.3	0.90	43.4	27.5	0.63	1846
AC34245	99	69	0.70	34.1	30.3	0.89	44.7	27.5	0.62	2091
AC34280	99	50	0.51	32.9	31.6	0.96	48.8	24.8	0.51	1580

for one to two weeks based on their tolerance level. Hence with a view to study the effect of submergence on germination and seedling attributes, eight cultivated varieties were subjected to submergence tolerance at five levels of flooding, viz. 1 cm, 2 cm, 3 cm, 4 cm and 5 cm flooding levels. CR 1009 Sub 1, a submergence-tolerant variety was used as check. With increase in submergence levels, greater reduction was observed for all the parameters. Survival percentage and seedling length were found to decrease under flooded conditions, but to a much lower extent in the tolerant genotype. Flooding decreased shoot, root and total dry matter production in all the varieties with more reduction in higher flooding (5 cm) rather than lower flooding (1 cm) levels (Fig. 3.5).

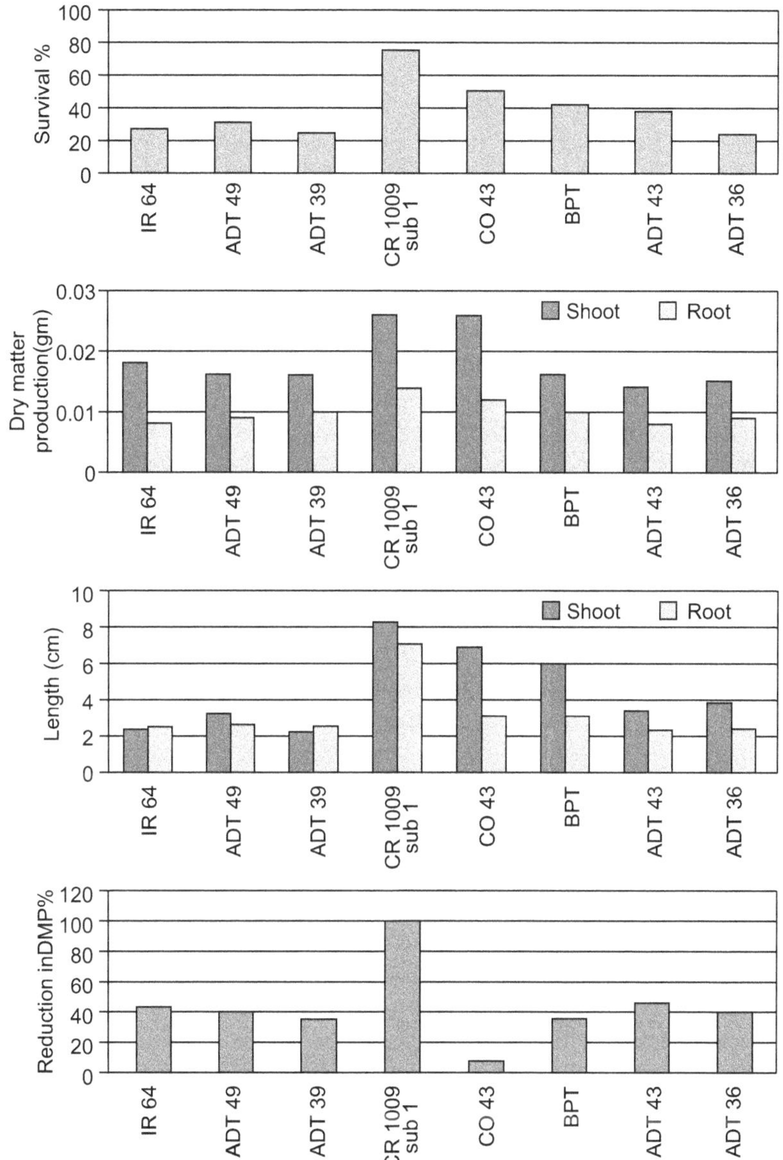

Fig. 3.5: Rice seed germination and seedling parameters.

The germination and early seedling growth stage of rice has been found to be highly intolerant to submergence (Ismail et al., 2009; Angaji et al., 2010; Joshi et al., 2013). Apart from germination, submergence also brings many morphological and physiological changes in rice plant. During submergence, rice plant survives by elongation of leaf sheath and blade at seedling stage and internodes at vegetative growth stage (Haque, 1974). If the flooding duration exceeds two to three weeks, even the submergence-tolerant varieties try to expose their leaf tip above the water surface for ensuring survival (Sarkar and Bhattacharjee, 2011; Colmer et al., 2014). This elongation process is an energy-extensive process, which curtails the availability of energy required for post-submergence plant growth, thus adversely affecting the post-submergence survival (Singh et al., 2001; Das et al., 2005).

Rice coleoptiles can elongate faster at low O_2 concentrations than in air, but germinating seeds normally fail to form leaves and roots (Alpi and Beevers, 1983; Ishizawa and Esashi, 1984). However, variation in the rate of coleoptile elongation under low oxygen (0–5%) was observed among 10 rice genotypes with maximum elongation observed at 0% oxygen (Turner et al., 1981). Rapid elongation of coleoptiles could facilitate contact with air in waterlogged or flooded soils and subsequent aeration of the growing embryo. Traits associated with coleoptile elongation of pre-germinated seeds under anoxia have been investigated before, and this growth was found to be independent of ethylene synthesis (Pearce and Jackson, 1991; Pearce et al., 1992), but dependent on the rate of ethanol synthesis (Setter et al., 1994), suggesting the importance of anaerobic metabolism during germination in flooded soils.

A promising strategy to increase seedling emergence and establishment is to breed rice varieties with rapid seed germination and coleoptile elongation, but progress has been slow (Fig. 3.7). The development of rice varieties that can germinate and emerge in flooded soils will therefore help reduce the hazards of early floods, to which rice is very sensitive (Ismail et al., 2009), and this also provides an efficient means for weed control through early flooding (El-Hendawy et al., 2011, 2012). These AG lines are tolerant to anaerobic conditions during germination (Septiningsih et al., 2013). The seeds of most AG lines absorb water rapidly from the start of germination; therefore, these genotypes germinate rapidly under anaerobic conditions (El-Hendawy et al., 2011). However, the physiological basis underlying the mechanism of rapid seed germination in AG lines has not been sufficiently described (Fig. 3.6).

Submergence during germination imposes germination stage oxygen deficiency (GSOD). Natural variation in anaerobic germination potential (AGP), manifested by their differences in GSOD tolerance, is available in rice. Present study was performed to understand differential responses of two rice cultivars, Naveen and AC41620 (susceptible and tolerant to GSOD, respectively), subjecting them to 48-hours of continuous submergence during germination. Transcriptome analysis revealed more elaborate regulation of gene expression in AC41620 compared to Naveen. Validation of transcriptome data through q-PCR of selected genes and biochemical

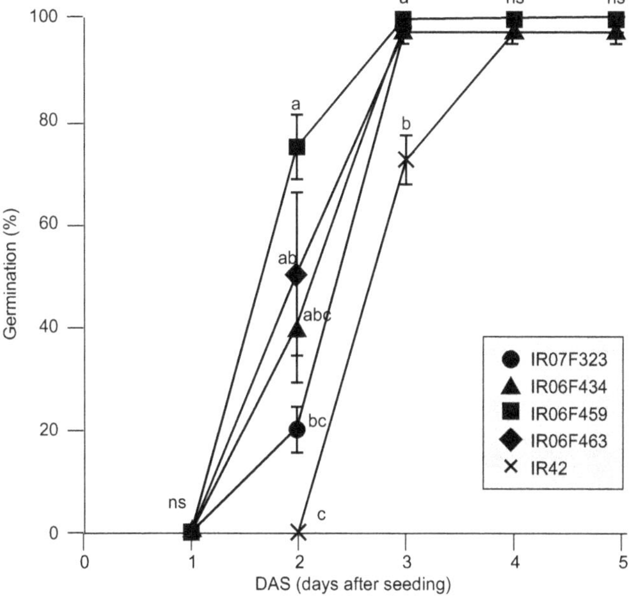

Fig. 3.6: Rice seed germination curves under submergence.

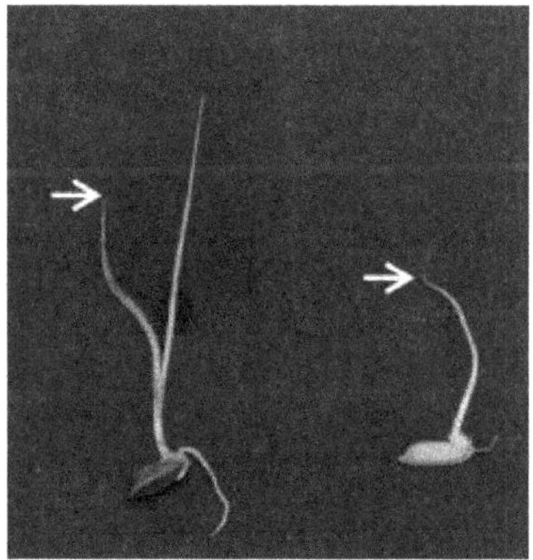

Fig. 3.7: Difference of coleoptile elongation under submergence (*left*) and normal (*right*) conditions.

analysis related to carbohydrate metabolism, anaerobic respiration and oxidative stress tolerance were more significant in AC41620 than Naveen.

Floods have destroyed over 2,000 acres of rice in Uganda which affected yield and caused losses to farmers. This problem is more pronounced when fields are not well levelled, and the mode of irrigation is by surface flooding. Majority of lowland rice fields in East African region are of this nature and are thus prone to yield losses. There are no submergence-tolerance varieties identified in Uganda so far.

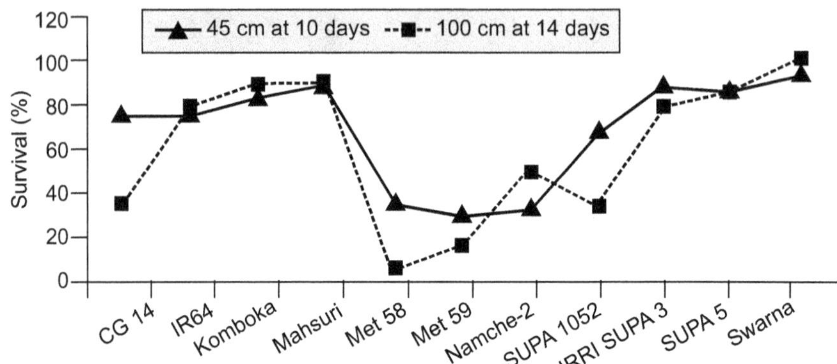

Fig. 3.8: Survival per cent of rice varieties at 10 and 14 days.

To address this problem, breeding for submergence tolerance is the most ideal and promising strategy in rice. As a first step, genotypes tolerant to submergence need to be identified which is the objective of this study. As many as 29 rice genotypes were morphological characterized in screen house and field conditions while 34 rice genotypes were molecularly characterized. Results suggested significant differences in the performance of genotypes both in the screen house and under field conditions in which varieties Swarna, IRRI SUPA 3 and KOMBOKA showed approximately 80% and above survival rate with Swarna variety ranking first. Molecular characterization of rice genotypes revealed that out of 34 genotypes, 30 genotypes scored presence for Sub 1A-2 allele while, four genotypes were neither Sub1A-1 nor Sub 1A-2 alleles. None of the tested genotypes were carrying Sub 1A-1 allele (Fig. 3.8).

Tillering in rice (*Oryza sativa* L.) is an important agronomic trait for grain production, and also a model system for the study of branching in monocotyledonous plants. Rice tiller is a specialized grain-bearing branch that is formed on the unelongated basal internode and grows independently of the mother stem (culm) by means of its own adventitious roots. Rice tillering occurs in a two-stage process: the formation of an axillary bud at each leaf axil and its subsequent outgrowth. Although the morphology and histology and some mutants of rice tillering have been well described, the molecular mechanism of rice tillering remains to be elucidated. It was reported that isolation and characterization of MONOCULM 1 (MOC1), a gene that is important in the control of rice tillering. The MOC1 mutant plants have only a main culm without any tillers owing to a defect in the formation of tiller buds. MOC1 encodes a putative GRAS family nuclear protein that is expressed mainly in the axillary buds and functions to initiate axillary buds and to promote outgrowth.

In wild-type rice plants, a tiller bud is normally formed at each leaf axil, but only those formed on the unelongated basal internodes can

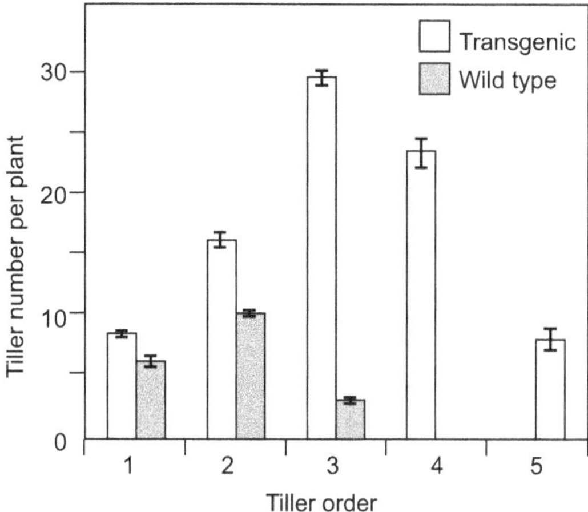

Fig. 3.9: Tiller number per plant in transgenic and wild type rice.

Table 3.2: Effects of submergence on rice yield, yield components, and effective leaf area in plants at the tillering stage.

Depth-Duration	Panicles/Pot	Seeds per Panicles	Seed Setting Rate (%)	1000 Grain Weight (g)	Yield (g/Pot)	High-tilling Panicles/Pot	P_o Rate (μmol $cm^{-2}s^{-1}$)	Photo-effective Leaf Area (cm^2/stem)
CK	27.1bc	105.6a	91.1bc	32.5a	84.6b	0.0d	27.9ab	83.1b
½-1	26.8bc	99.1b	93.4a	32.1a	79.5c	0.0d	27.2b	77.0c
½-3	27.5bc	98.4b	88.9de	32.0a	77.0d	0.3cd	26.5bc	74.7cd
½-5	27.3bc	99.2b	86.3f	32.2a	75.2d	1.3bc	26.1bc	73.7cd
2/3-1	26.3c	98.0b	92.8ab	32.1a	76.8d	0.3cd	27.7ab	87.8ab
2/3-3	27.2bc	94.3c	90.3cd	32.5a	75.3d	0.5cd	25.4c	74.6cd
2/3-5	27.8bc	93.2c	88.1ef	31.7a	72.3e	1.7b	25.0c	73.3cd
1/1-1	28.2b	105.8a	93.4a	32.8a	91.3a	0.5cd	28.6a	89.6a
1/1-3	31.8a	88.0d	88.0ef	32.1a	79.1c	1.2bc	24.8c	72.7cd
1/1-5	32.0a	87.5d	87.7ef	31.3a	76.9d	2.8a	23.4d	71.4d

grow out into tillers and those formed on the elongated upper internodes become arrested (Fig. 3.9). Secondary tillers are usually formed in wild-type plants, but higher-order tillers, such as tertiary, quaternary and quinternary ones, are seldom developed.

Flooding is a major threat to agricultural production. Most studies have focused on the lower water-storage limit in rice fields, whereas few studies have examined the upper water-storage limit. This study aimed to explore the effect of waterlogging at the rice tillering stage on rice growth and yield. The early-ripening late japonica variety, Yangjing 4227, was selected for this study. The treatments included different submergence depths (submergence depth/ plant height: 1/2 (waist submergence), 2/3 (neck submergence), and 1/1 (complete submergence)) and durations (1, 3, and 5 days). The control group was treated with conventional alternation of drying and wetting. The effects of waterlogging at the tillering stage on root characteristics, dry matter production, nitrogen and phosphorus accumulation, yield, yield components, and 1-aminocyclopropane-1-carboxylic acid synthase (ACS) gene expression were explored. Compared with the control group, the 1/1 group showed significant increases in yield, seed-setting rate, photosynthetically efficient leaf area, and OS-ACS3 gene expression after one day of submergence. The grain number per panicle, dry weight of the above-ground and below-ground parts, and number of adventitious roots also increased. Correlation analysis revealed a significant positive correlation between the panicle number and nitrogen content; however, no significant correlation was found for phosphorus content. If a decrease in rice yield of less than 10% is acceptable, half, two-third, and complete submergence of the plants can be performed at the tillering stage for one to three days; this treatment will increase the space available for rice field water management/control and will improve rainfall resource utilization.

Values followed by a different letter are significantly different (t-test, $P < 0.05$). The main strategies enabling rice plants to cope up with flash flooding stress require growth regulation during submergence and subsequent rapid growth recovery after de-submergence and better nutrient management options can enhance this strategy. Application methods of nitrogen and phosphorus were evaluated in submerged rice for tiller mortality, productivity, grain quality and nutrient absorption. The performance of Sub1 (IR-64 Sub1 and Swarna Sub1) and non-Sub1 (IR-20) cultivars of rice was tested under clear and turbid water for their tolerance to submergence in response to basal phosphorus and post-submergence nitrogen. Tillering ability, yield and nutrient absorption of rice subjected to complete submergence for 15 days decreased significantly over non-submerged rice plants. Turbid water submergence was fatal in terms of tiller mortality, reduced nutrient absorption and yield because of low light and dissolved oxygen underwater. The crop fertilized with nitrogen produced a greater number of tillers, yield attributes and grain yield than the unfertilized crop.

Basal Si application promoted rice growth and development at the tillering stage, improving rice seedling tolerance to submergence stress. After the occurrence of flooding, timely foliar spraying of N or Si had significant remedial effects, and spraying N along with Si had a better effect. The combination of basal Si and post-flooding N and Si spraying was the most promising method of nutrient application and as resistance to damage during submergence was enhanced, plants rapidly resumed growth and development. By maintaining a great number of green leaves and tillers as well as higher above-ground and below-ground dry mass, sugar contents and antioxidant enzyme activities, these plants yielded significantly more grain.

An experiment was performed to evaluate the impact of zinc sulfate by examining survival rate, growth attributes, and biochemical parameters under the submerged and non-submerged conditions to enhance submergence tolerance in Sub1 and non-Sub1 rice varieties. A fully randomized block design was followed with three treatments, three replications, and five chosen rice varieties centered on submergence resistance and susceptibility. Sowing seeds directly with the treatments comprises viz. T1: 0.0 mg $ZnSO_4$ per kg of soil, T2: 5.0 mg $ZnSO_4$ per kg, T3: 10 mg $ZnSO_4$ per kg of soil. Treatments were performed in both submerged and non-submerged (controlled) pots at sowing. After 30 days of sowing, 15 days of full submergence, it follows that zinc sulphate @10 mg/kg soil greatly improved survival percentage, growth qualities, and biochemical parameters of Sub1 and nonSub1 rice varieties even after 15 days of total submergence.

3.9 Submergence, Depth and Duration

Complete plant submergence for six or nine days at 20 days after transplanting effected the same decrease in grain yield as submergence for 12 days at 40 days after transplanting. With increasing duration of submergence, tiller number, green leaves and dry weight of all varieties that were tested decreased. The decrease was less in the flood-tolerant variety FR 13A than in other varieties. Contents of reducing sugars and amylase activity also decreased with increasing duration of submergence. The reducing sugar contents and amylase activity were higher and peroxidase activity was lower in flood-tolerant variety FR 13A than in other varieties. The N contents increased and P and K contents decreased with duration of submergence.

Longer-term partial or stagnant flooding is also common in low-lying areas, where water accumulates throughout the growing season and at various depths (Mackill et al., 1996; Singh et al., 2011). A water depth of 30–50 cm is common in flood-affected rain-fed areas, and is referred to as medium-deep or stagnant floods (SF). These water depths depress yields by hindering tillering and plant growth and can result in lodging and poor grain quality. Under more extreme conditions, water depths can reach several metres, and this is referred to as deepwater or floating.

Evaluation of 18 rice cultures for growth and yield attributes in waterlogged fields was conducted in 1983 in the Gangetic plains of West Bengal in *kharif* season. Eight varieties had better than 70% survival.

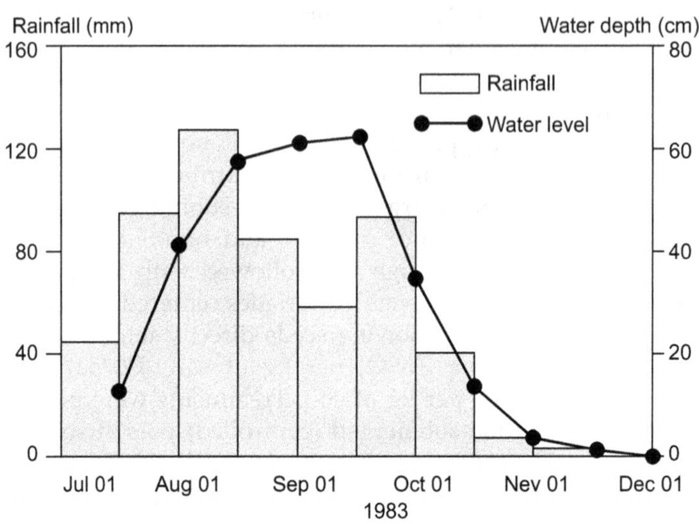

Fig. 3.10: Rainfall (mm) and water depth (cm) variations during the experiment.

Days to 50% flowering ranged from 118 to 150. OR330-40-3 flowered earliest and CR292-5258 and CR292-7050 flowered last. Total dry matter production varied widely. BIET807 accumulated 1,308 g dry weight per m^2 and PLA1100 dry weight of 242 g per m^2. Survival percentage appeared to substantially influence dry matter production irrespective of days to 50% flowering or maturity. Tilakkachari produced 255 panicles per m^2, followed by NC492 253 panicles. Tilakkachari also produced the maximum spikelets. NC492 yielded highest with 4.9 t/ha and PLA1100 lowest with 0.80 t/ha. NC492 had high dry matter, high panicles number, the highest 1,000 grain weight and high survival percentage. Survival percentage of plants under waterlogged situation is the prime criterion for plant establishment and subsequent manifestation of yield potential (Fig. 3.10 and Table 3.3) (Basuchaudhury and DasGupta, 1985).

Submergence stress can greatly limit grain yield of inbred rice (*Oryza sativa* L.), but the effects of submergence on hybrid rice are unclear. A pot experiment was conducted to clarify the effects of submergence that happens at the vegetative growth stage on two Chinese hybrid rice cultivars. The rice cultivars Zheyou-18, an indica-japonica hybrid, and Yliangyou-689, an indica hybrid, were planted under two treatments, submergence (43 cm depth of tap water for about two weeks) and control (no submergence). Results showed that the grain yield of submerged 'Yliangyou-689' was 539.0 g m^{-2} in 2018 and 614.2 g m^{-2} in 2019; submergence significantly reduced grain yield by 49.6% and 44.2% of that

Table 3.3: Survival, growth, and yield attributes of rice varieties under waterlogged condition.

Variety	Survival (%)	Grain Yield (t/ha)	Days to 50% Fowering	Dry Matter Production (g/m^2)	Panicles/m^2	Spikelets (10^3/m^2)	Grains (10^3/m^2)	1000-Grains Wt. (g)
OR143-7	36	2.0	141	766	117	15	12	27
OR117-215	17	2.2	132	600	50	12	10	21
OR330-40-3	80	2.2	118	758	154	17	14	21
Janki	87	2.1	122	875	200	15	13	21
BIET724	70	3.1	135	1002	172	20	15	26
BIET807	80	3.6	125	1308	200	23	18	24
BIET821	34	2.0	128	725	119	17	11	24
CR292-5258	31	2.2	150	489	144	20	15	18
CR292-7050	21	2.0	150	466	69	8	7	30
CR221-1030	87	3.5	134	792	182	22	19	22
CN499-160-2-1	29	2.2	134	617	126	12	09	18
CN499-160-13-6	83	2.1	135	608	200	21	17	17
CN505-5-32-9	12	1.2	140	325	50	11	8	20
CN506-147-14-2	14	3.3	135	685	90	19	16	23
CN506-147-2-1	93	2.4	142	880	200	19	15	25
PLA1100	12	0.8	144	242	27	7	5	20
NC492	90	4.9	130	1116	253	21	19	31
Tilakkachari	93	3.8	132	1205	255	25	17	26
Mean	53.8	2.5	134.8	747.7	144.9	16.99	13.4	23.02
CDat5%P	21	0.6	4	200	52	3	3	3

Table 3.4: Grain yields and yield components of two rice cultivars subjected to submergence or not (control).

Variety	Treatment	Grain Yield (g/m^2)	Spikelets/m^2	Panicles/m^2	Spikelets/ panicle	Grain Filling (%)	Grain Wt. (mg)
2018 Yliangyou-689	Control	1069.9a	52708a	296.1a	182.2a	77.9a	22.4a
	Submer gence	539.0b	34145b	133.8b	256.7a	66.7b	20.4b
Zheyou-18	control	661.9a	42211a	191.0a	220.2a	67.1b	20.2a
	Submer gence	466.8a	20998b	148.0a	144.3b	88.5a	21.7a
2019 Yliangyou-689	Control	1101.8a	66571a	390.4a	169.7a	70.9a	20.2a
	Submer gence	614.2b	34886b	261.9b	187.4b	72.7a	20.9a
Zheyou-18	Control	521.5a	55205a	223.9a	249.8a	40.8a	20.0a
	Submer gence	376.4a	44524a	161.9a	275.4a	37.5a	19.3a
ANOVA	Year(Y)	NS	**	*	NS	**	NS
	Treat(T)	**	**	**	NS	NS	NS
	Var(V)	**	NS	**	NS	NS	NS
	YxT	NS	NS	NS	NS	NS	NS
	VxT	**	NS	NS	NS	NS	NS

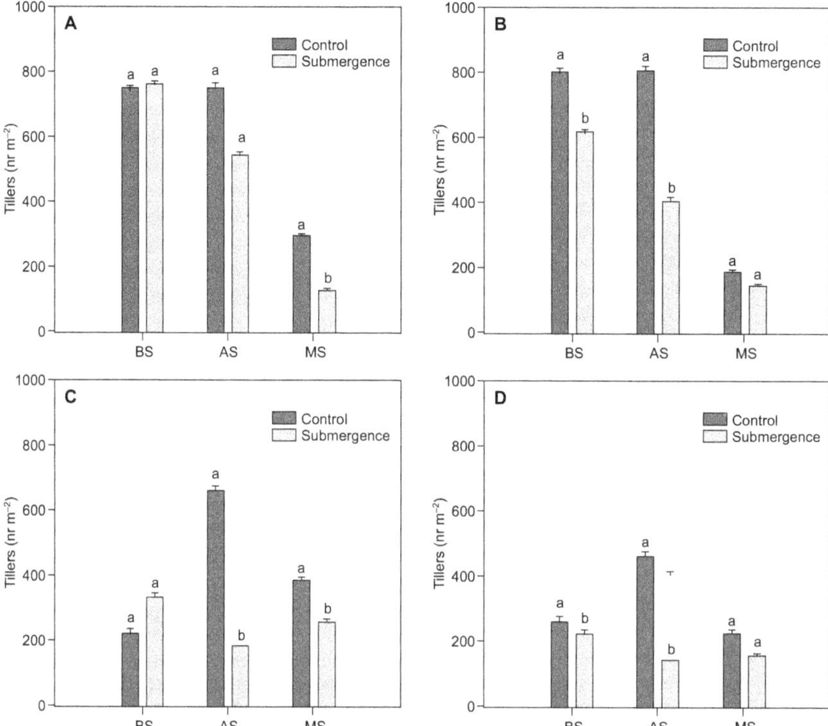

Fig. 3.11: Tillering performances under control and submergence at different times.

of the control. The lower level of grain yield was attributed to reduced survival rate, number of tillers and spikelets, plant dry weight, and crop growth rate, and excessive elongation of stems due to submergence of plants. The grain yield of submerged 'Zheyou-18' was 466.8 g m^{-2} in 2018 and 376.4 g m^{-2} in 2019 and neither of which were significantly different from that of the control, indicating higher submergence tolerance in 'Zheyou-18' than in 'Yliangyou-689'. Submergence did not affect plant N and K contents, but it reduced plant N uptake rates by 47.8% to 88.7% and it reduced K uptake rates by 53.9% to 89.5% of that of the control in these two hybrid cultivars. Furthermore, non-significant differences were observed in all of the parameters related to rice starch viscosity between control and submergence treatment (Table 3.4) (Chen et al., 2021).

Ten rice genotypes including two submergence-tolerant checks, two susceptible varieties and six advanced lines were evaluated for submergence tolerance in the laboratory and in the field during January–December 2015. The experiment was conducted in the field following

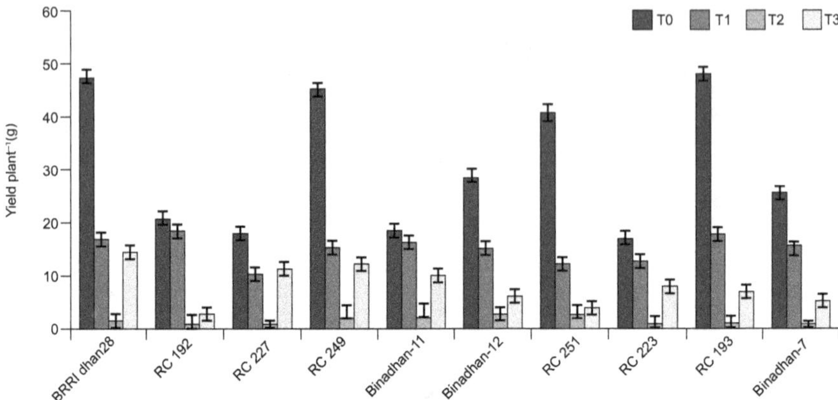

Fig. 3.12: Yield per plant of some rice varieties under different submergence treatments.

randomized complete block design in a two-factor arrangement using five replications. Ten characters, viz. days to flowering, plant height, tiller number plant^{-1}, effective tiller plant^{-1}, and yield plant^{-1}, etc., were studied for four treatments. A significant genotype environment interaction was observed for all traits studied in this experiment. The yield was reduced for all genotypes at a different level of submergence stress compared to control. Binadhan-11, Binadhan-12, RC 249 and RC 251 showed tolerance, whereas RC 192, RC 193 and RC 225 showed moderate tolerance in submerged condition. The phenotypic coefficient of variance (PCV) was higher than the genotypic coefficient of variance (GCV) in all the studied traits. High heritability (75–97%) was found for all traits. High heritability along with high genetic advance was found for days to flowering (45.55) and plant height (40.05). Molecular characterization of the used genotypes was done with three SSR markers, viz. RM 24, and submergence specific SC3 and SUB1. SC3 was found reliable for detection of submergence-tolerant genotypes due to the highest gene diversity (0.840) compared to others. The banding pattern of the submergence specific markers SC3 and SUB1 was identified in Binadhan-11, Binadhan-12, RC 192, RC 193, RC 225, RC 227, RC 249, and RC 251, which possess the SUB1 gene. Finally, clustering also separates the tolerant genotypes from the susceptible by dividing them into different clusters. The identified genotypes might be useful for the breeding programme for the development of submergence tolerant as well as resistant rice variety in Bangladesh (Arfin et al., 2018).

A study was conducted to evaluate the genotypes having Sub1 gene under submergence and drought stress in field conditions at Nuclear Institute for Agriculture and Biology (NIAB), Faisalabad. A pot experiment was also performed to study the influence of complete submergence on elongation and survival percentage of five Sub1 genotypes along with two high-yielding local cultivars. All Sub1 genotypes showed less elongation percentages than Super Basmati and KSK-133 while maximum survival percentage was observed in IR-07-F289 Sub1 followed by Swarna Sub1 and FR-13A. Five rice genotypes (FR-13A, Swarna Sub1, Ciherang Sub1, IR-07-F289 Sub1 and IR-44 Sub1) along with two high-yielding local varieties (KSK-133 and Super Basmati) were evaluated under submergence stress in field conditions. The field experiments were laid out in split plot randomized complete block design (RCBD) with three replications. Results of ANOVA revealed that chlorophyll contents, plant height, number of productive tillers per plant, panicle length, total spikelets per panicle, number of grains per panicle and 1,000 grain weight were affected by submergence stress. While in another experiment drought stress was applied for 30 days on five Sub1 genotypes along with Nagina-22 (Drought tolerant check) and IR-64 (drought susceptible check) in split plot design with three replications. Drought stress severely reduced all the parameters under study, except leaf area, number of productive tillers per plant and biological yield per plant which remain unaffected. Overall results revealed FR-13A only produced grains under submergence stress while under normal and drought conditions, it did not produce grains. All Sub1 genotypes performed well under submergence stress. Swarna Sub1 significantly produced more primary branches per panicle (10.5), yield (5.13 g) and harvest index (13.09) under drought stress as compared to the Nagina-22. All the remaining Sub1 genotypes also showed better performance than drought susceptible check (IR-64) and showed non-significant difference with Nagina-22 for most of the drought-tolerance related traits (Table 3.5) (Amjad et al., 2022).

Table 3.5: Elongation per day, elongation % and survival % of rice genotype under complete submergence.

S. No.	Variety	Plant ht. (cm) Before Stress	Plant ht.(cm) After Stress	Elongation/day (cm)	Elongation %	Survival %
1	FR13A-Sub1	32.3	49.5	0.95	53.25	55.56
2	Swarna Sub1	22.57	26.5	0.21	17.41	66.67
3	IR-44 Sub1	26.94	30.33	0.18	12.58	33.33
4	IR-07-F 289Sub1	30.04	44.83	0.182	49.23	77.78
5	Ciherang Sub1	30.43	32.66	0.123	7.32	33.33
6	KSK-133	30.77	56.6	1.44	83.94	55.56
7	Super Basmati	29.66	52.5	1.27	77.00	44.40

Table 3.6: Effect of submergence on yield of hybrid rice.

Treatment	Yield/m^2 (kg)	Reduction of Yield (%)
D0	0.62 a	0.00 d
D0	0.60 a	5.23 c
D0	0.50 b	19.02 b
D0	0.38 c	31.03 a
D0	6.88	7.89

D0 = No submergence, D1 = Six days submergence, D2 = Ten days submergence and D3 = Fourteen days submergence

Improvement of combined tolerance to submergence and drought would substantially increase rice productivity. Results showed that all Sub1 genotypes performed well not only under submergence stress but also under drought stress specially, Swarna Sub1.

The whole field was divided into three equal blocks, each containing 24 plots. Each block was subdivided into four sub-blocks. As such, there were 12 sub-blocks. Each sub-block was encircled by the 50 cm high soil wall ridge, which was hundred per cent water leakage-proof. In total, there were 72 plots. BRRI dhan51 followed by BRRI hybrid dhan1 showed lower grain yield reduction percentage in submerged conditions compared to control by attaining good yield contributing characters and thus proved as tolerant varieties. On the other hand, BRRI dhan34 and ACI hybrid1 were susceptible to submergence (Table 3.6) (Sultana et al., 2018).

References

Alpi, A., and Beevers, H. (1983). Effects of O_2 concentration on rice seedlings. *Plant Physiology,* 71: 30–34.

Amjad, I., Khalid, M., Kashif, M., Norman, M., Ali, S., Shaikh, R. et al. (2022). Performance evaluation of Sub1 rice genotypes for vegetative stage submergence stress and reproductive stage drought stress. *Sarhad J. Agric.*, 38: 240–251.

Angaji, S.A., Septiningsih, E.M., Mackill, D.J., and Ismail, A.M. (2010). QTLs associated with tolerance of flooding during germination in rice (*Oryza sativa* L.), *Euphytica,* 172: 159–168.

Afrin, W., Nafis, M.H., Hossain, M.A., Islam, M.M., and Hossain, M.A. (2018). Responses of rice (*Oryza sativa* L.) genotypes to different levels of submergence, *C R. Biol.*, 341(2): 85–96.

Armstrong, W., and Drew, M.C. (2002). Root growth and metabolism under oxygen deficiency. pp. 729–761. *In*: Waisel, Y., Eshel, A., Beeckman, T. and Kafkafi, U. (Eds.). *Plant Roots: The Hidden Half,* 3rd edition, New York, Basel.

Ashraf, M.A., Ahmad, M.S., Ashraf, M., Al-Qurainy, F. and Ashraf, M.Y. (2011). Alleviation of waterlogging stress in upland cotton (*Gossypium hirsutum* L.) by exogenous application of potassium in soil and as a foliar spray. *Crop Pasture Sci.*, 62: 25–38.

Ayano, M., Kani, T., Kojima, M., Sakakibara, H., Kitaoka, T., Kuroha, T. et al (2014). Gibberellin biosynthesis and signal transduction is essential for internode elongation in deepwater rice, *Plant Cell Environ.*, 37: 2313–2324.

Ayi, Q., Zeng, B., Liu, J., Li, S., van Bodegom, P.M., and Cornelissen, J.H.C. (2016). Oxygen absorption by adventitious roots promotes the survival of completely submerged terrestrial plants. *Ann. Bot.*, 118: 675–683.

Azarin, K.V., Usatov, A.V., and Kostylev, P.I. (2017). Molecular breeding of submergence-tolerant rice, *Annu. Res. Rev. Biol.*, 18(1): 1–10.

Bailey-Serres, J., and Voesenek, L.A.C.J. (2008). Flooding stress: Acclimations and genetic diversity. *Annu. Rev. Plant Biol.*, 59: 313–339.

Bailey-Serres, J., Fukao, T., Ronald, P.C., Ismail, A. M., Heuer, S., and Mackill, D.J. (2010). Submergence tolerant rice: *Sub1*'s journey from landrace to modern cultivar. *Rice*, 3: 138–147.

Bailey-Serres, J., Fukao, T., Gibbs, D.J., Holdsworth, M.J., Lee, S.C., Licausi, F. et al. (2012). Making sense of low oxygen sensing. *Trends in Plant Science*, 17: 129–138.

Balasubramanian, V., and Hill, J. (2002). Direct seeding of rice in Asia: Emerging issues and strategic research needs for 21st century. *In: Direct Seeding Research Strategies and Opportunities*, pp. 15–39.

Basuchaudhury, P., and DasGupta, D.K. (1985). NC492 is a promising rice variety for waterlogged fields of West Bengal. *International Rice Research Newsletter*, 10(2).

Bhaduri, D., Chakraborty, K., Nayak, A.K., Shahid, M., Tripathi, R., Behera, R. et al. (2020). Alteration in plant spacing improves submergence tolerance in Sub1 and non-Sub1 rice (cv. IR64) by better light interception and effective carbohydrate utilization under stress. *Funct. Plant Biol.*, 47(10): 891–903.

Bin Rahman, A.N.M.R., and Zhang, J.H. (2016). Flood and drought tolerance in rice: Opposite but may coexist. *Food Energy Secur.*, 5(2): 76–88.

Bishnoi, U., Jain, R.K., Rohilla, J.S., Chowdhury, V.K., Gupta, K.R. and Chowdhury, J.B.(2000). Anther culture of recalcitrant indica × Basmati rice hybrids. *Euphytica*, 114: 93–101.

Catling, D. (1992). *Rice in Deep Water*. London: MacMillan Press.

Chamara, B.S., Marambe, B., Kumar, V., Ismail, A.M., Septiningsih, E.M., and Chauhan, B.S. (2018). Optimizing sowing and flooding depth for anaerobic germination-tolerant genotypes to enhance crop establishment, early growth, and weed management in dry-seeded rice (*Oryza sativa* L.). *Front. Plant Sci.*, 9: 1654.

Chen, Y., Song, J., Yan, C., and Hong, X. (2021). Effects of submergence stress at the vegetative growth stage on hybrid rice growth and grain yield in China. *Chilean Journal of Agricultural Research*, 81(2): 191–201.

Collard, B.C.Y., Septiningsih, E.M., Das, S.R., Carandang, J.J., Sanchez, D.L., Kato, Y. et al. (2013). Developing new flood-tolerant varieties at the International Rice Research Institute (IRRI) towards 2025. *SABRAO Journal of Breeding and Genetics*, 45: 42–56.

Colmer, T.D., and Flowers, T.J. (2008). Flooding tolerance in halophytes. *New Phytologist*, 179: 964–974.

Colmer, T.D., and Voesenek, L.A.C.J. (2009). Flooding tolerance: Suites of plant traits in variable environments. *Funct. Plant Biol.*, 36(8): 665–681.

Colmer, T.D., Winkel, A., and Pedersen, O. (2011). A perspective on underwater photosynthesis in submerged terrestrial wetland plants. *AoB Plants*, 1–12.

Colmer, T.D., Armstrong, W., Greenway, H., Ismail, A.M., Kirk, G.J.D., and Atwell, B.J. (2014). Physiological mechanisms in flooding tolerance of rice: Transient complete submergence and prolonged standing water. *Progress in Botany*, 75: 255–307.

Crawford, R.M.M., and Braendle, R. (1996). Oxygen deprivation stress in a changing environment. *J. Exp. Bot.*, 47: 145–159.

Dar, M.H., Janvry, A. d.e, Raitzer, K.D., and Sadoulet, E. (2013). Flood-tolerant rice reduces yield variability and raises expected yield , differentially benefitting socially disadvantaged groups. *Sci.Reports,* 3: 3315, doi: 10.1038 /srep03315.

Dar, M.H., Chakravorty, R., Waza, S.A., Sharma, M., Zaidi, N.W., Singh, A.N. et al. (2017). Transforming rice cultivation in flood-prone coastal Odisha to ensure food and economic security. *Food Secur.,* 9: 711–722.

Das, K.K., Panda, D., Nagaraju, M., Sharma, S.G., and Sarkar, R.K. (2004). Antioxidant enzymes and aldehyde releasing capacity of rice cultivars (*Oryza sativa* L.) as determinants of anaerobic seedling establishment capacity. *Bulg. J. Plant Physiol.,* 30: 34–44.

Das, K.K., Sarkar, R.K., and Ismail, A.M. (2005). Elongation ability and non-structural carbohydrate levels in relation to submergence tolerance in rice. *Plant Sci.,* 168: 131–136.

Das, K.K., Panda, D., Sarkar, R.K., Reddy, J.N., and Ismail, A.M. (2009). Submergence tolerance in relation to variable floodwater conditions in rice. *Environ. Exp. Bot.,* 66: 425–434.

Drew, M.C., and Lynch, J.M. (1980). Soil anaerobiosis, microorganisms, and root function. *Annual Review of Phytopathology,* 18: 37–66.

Du, L.V., and Tuong T.P. (2002). Enhancing the performance of dry seeded rice: Effects of seed priming, seedling rate, and time of seedling. pp. 241–256. *In:* Pandey. S., Mortimer, M., Wade, L., Tuong, T.P., Lopez, K., and Hardy, B. (Eds.). *Direct Seeding: Research Strategies and Opportunities,* 196 Rice Science, Vol. 22, No. 4, Manila, the Philippines: International Rice Research Institute.

El-Hendawy, S.E., Sone, C., Ito, O., and Sakagami, J.I. (2011). Evaluation of germination ability in rice seeds under anaerobic conditions by cluster analysis. *Res. J. Seed Sci.,* 4: 82–93.

El-Hendawy, S.E., Sone, C., Ito, O., and Sakagami, J.I. (2012). Differential growth response of rice genotypes based on quiescence mechanism under flash flooding stress. *Aust. J. Crop Sci.,* 6: 1587–1597.

Ella, E.S., Kawano, N., Yamauchi, Y., Tanaka, K., and Ismail, A.M. (2003). Blocking ethylene perception during submergence reduced chlorophyll degradation and improved seedling survival in rice. *Funct. Plant Biol.,* 30: 813–819.

Ella, E.S., Dionisio-Sese,M.L., and Ismail, A.M. (2011). Seed pre-treatment in rice reduces damage, enhances carbohydrate mobilization and improves emergence and seedling establishment under flooded conditions. *AoB Plants,* 2011:plr007. Doi: 10.1093/aobpla/plr007.

Farooq, M., Wahid, A., Kobayashi, N., Fujita, D., and Basra, S.M.A. (2009). Plant drought stress: effects, mechanisms and management. *Agron. Sustain. Dev.,* 29: 185–212.

Fleck, A.T., Nye, T., Repenning, C., Stahl, F., Zahn, M., and Schenk, M.K. (2011). Silicon enhances suberization and lignification in roots of rice (*Oryza sativa*). *Journal of Experimental Botany,* 62: 2001–2011.

Fukao, T., Xu, K., Ronald, P.C., and Bailey-Serres, J. (2006). A variable cluster of ethylene response factor-like genes regulates metabolic and developmental acclimation responses to submergence in rice. *Plant Cell,* 18(8): 2021–2034.

Fukao, T., and Bailey-Serres, J. (2008). Submergence tolerance conferred by Sub1A is mediated by SLR1 and SLRL1 restriction of gibberellin responses in rice. *Proc. Natl. Acad. Sci.* USA., 105(43): 16814–16819.

Gautam, P., Lal, B., Raja, R., Baig, M.J., Haldar, D., Rath, L. et al. (2014). Post–flood nitrogen and basal phosphorus management affects survival, metabolic changes and anti-oxidant enzyme activities of submerged rice (*Oryza sativa*). *Functional Plant Biology,* 41: 1284–1294.

Hancock, R.D., and Viola, R. (2005). Biosynthesis and catabolism of L-ascorbic acid in plants. *Crit. Rev. Plant Sci.,* 24: 167–188.

Haque, M.Z. (1974). *Physiological behaviour of deep water rice.* Deep water rice in Bangladesh. Bangladesh Rice Research Institute. pp. 45–59.

Hattori, Y., Nagai, K., Furukawa, S., Song, X.J., Kawano, R., Sakakibara, H. et al. (2009). The ethylene response factors SNORKEL1 and SNORKEL2 allow rice to adapt to deep water. *Nature,* 460(7258): 1026–U1116.

Hattori, Y., Nagai, K., and Ashikari, M. (2011). Rice growth adapting to deepwater. *Curr. Opin. Plant Biol.,* 14(1): 100–105.

He, D., and Yang, P. (2013). Proteomics of rice seed germination. *Frontiers in Plant Science,* 4: 246.

Htwe, N.M., Singleton, G.R., and Johnson, D.E. (2019). Interactions between rodents and weeds in a lowland rice agro-ecosystem: The need for an integrated approach to management. *Integ. Zool.,* 14: 396–409.

Huke, R.E. and Huke, E.H. (1997). *Rice Area by Type of Culture, South, Southeast, and East Asia*: A Revised and Updated Database, International Rice Research Institute, Los Ban˜os, the Philippines.

Hussain, S., Khan, F., Hussain, H.A., and Nie, L. (2016a). Physiological and biochemical mechanisms of seed priming-induced chilling tolerance in rice cultivars. *Front. Plant Sci.,* 7: 116.

Hussain, S., Yin, H., Peng, S., Khan, F.A., Khan, F., Hussain, H.A. et al. (2016b). Comparative transcriptional profiling of primed and non-primed rice seedlings under submergence stress. *Front. Plant Sci.* 7: 1125.

Iftekharuddaula, K., Newaz, M., Salam, M., Ahmed, H., Mahbub, M., Septiningsih, E. et al. (2011). Rapid and high-precision marker assisted backcrossing to introgress *SUB1 QTL* into BR11, the rain-fed lowland rice mega variety of Bangladesh. *Euphytica,* 178: 83–97.

Ishizawa, K., and Esashi, Y. (1984). Gaseous factors involved in the enhanced elongation of rice coleoptiles under water. *Plant, Cell and Environment,* 7: 239–245.

Ismail, A.M., Ella, E.S., Vegara, G.V., and Mackill, D.J. (2008). Mechanisms associated with tolerance to flooding during germination and early seedling growth in rice (*Oryza sativa*). *Ann. Bot.,* 103: 197–209.

Ismail, A.M., Ella, E.S., Vergara, G.V., and Mackill, D.J. (2009). Mechanisms associated with tolerance to flooding during germination and early seedling growth in rice (*Oryza sativa*). *Ann. Bot.,* 103: 197–209.

Ismail, A.M., Johnson, D.E., Ella, E.S., Vergara, G.V., and Baltazar, A.M. (2012). Adaptation to flooding during emergence and seedling growth in rice and weeds, and implications for crop establishment. *AoB Plants,* 2012: pls019.

Ismail, A.M., Singh, U.S., Singh, S., Dar, M., and Mackill, D.J. (2013). The contribution of submergence-tolerant (Sub1) rice varieties to food security in flood-prone areas. *Field Crops Research,* 152: 83–93.

Jackson, M.B., and Ram, P.C. (2003). Physiological and molecular basis of susceptibility and tolerance of rice plants to complete submergence. *Ann. Bot.,* 91(2): 227–241.

Jisha, K.C., Vijayakumari, K., and Puthur, J.T. (2013). Seed priming for abiotic stress tolerance: an overview. *Acta Physiol. Plant,* 35: 1381–1396.

Joshi, R., Vig, A.P., and Singh, J. (2013). Vermicompost as soil supplement to enhance growth, yield and quality of *Triticum aestivum* L.: A field study. *Int, J, Recycl. Org. Waste Agric.,* 2(1): 16.

Kende, H., van der Knaap, E., and Cho, H.T. (1998). Deep-water rice: A model plant to study stem elongation. *Plant Physiol.,* 118: 1105–1110.

Khaliq, F., Ahmad, A.U.H., Farooq, M., and Murtaza, G. (2015). Evaluating the role of seed priming in improving the performance of nursery seedlings for system of rice intensification. *Pak. J. Agric. Sci.,* 52: 27–36.

Kuanar, S.R., Ray, A., Sethi, S.K., Chattopadhyay, K., and Sarkar, R.K. (2017). Physiological basis of stagnant flooding tolerance in rice. *Rice Sci.,* 24: 73–84.

Kumar, R., Mishra, J.S., Upadhyay, P.K., and Hans, H. (2019). Rice fallows in the eastern India: Problems and Prospects. *Indian J,.Agric. Sci.*, 89: 567–577.

Kuroha, T., Nagai, K., Gamuyao, R., Wang, D.R., Furuta, T., Nakamori, M. et al. (2018). Ethylene-gibberellin signalling underlies adaptation of rice to periodic flooding. *Science*, 361: 181–186.

Lee, K.W., Chen, P.W., and Yu, S.M. (2014). Metabolic adaptation to sugar/O2 deficiency for anaerobic germination and seedling growth in rice. *Plant Cell Environ.*, 37(10): 2234–2244.

Loreti, E., van Veen, H., and Perata, P. (2016). Plant responses to flooding stress. *Curr. Opin. Plant Biol.*, 33: 64–71.

Mamun, E.A., Alfred, S., Cantrill, L.C., Overall, R.L., and Sutton, B.G. (2006). Effects of chilling on male gametophyte development in rice. *Cell Biol. Int.*, 30: 583–591.

Mamun, M.A.A., Haque, M.M., Saleque, M.A., Khaliq, Q.A., Karim, M.A., and Karim, A.J.M.S. (2017). Nitrogen fertilizer management for tidal submergence tolerant landrace rice (*Oryza sativa* L.) cultivars. *Ann. Agric. Sci.* 62: 193–203.

Mackill, D.J., Coffman, W.R., and Garrity, D.P. (1996). *Rain-fed Lowland Rice Improvement*. International Rice Research Institute, Los Ban˜os, the Philippines.

Mackill, D.J. (2006). Breeding for resistance to abiotic stresses in rice: The value of quantitative trait loci. pp. 201–212. *In*: Arnel, T., Hallauer, R., International Symposium, Lamkey, K.R., and Lee, M. (Eds.). *Plant Breeding*, New York: Blackwell Publishing.

Mackill, D.J., Ismail, A.M., Singh, U.S., Labios, R.V., and Paris, T.R. (2012). Development and rapid adoption of submergence-tolerant (Sub1) rice varieties. *Adv Agron.*, 115: 299–352.

Magneschi, L., and Perata, P. (2009). Rice germination and seedling growth in the absence of oxygen. *Ann. Bot.*, 103(2): 181–196.

Manzanilla, D.O., Paris, T.R., Vergara, G.V., Ismail, A.M., Pandey, S., Labios, R.V. et al. (2011). Submergence risks and farmers' preferences: Implications for breeding Sub1 rice in Southeast Asia, *Agricultural Systems*. https://doi.org/10.1016/j.agsy.2010.12.005.

Metraux, J.P., and Kende, H. (1983). The role of ethylene in the growth response of submerged deep-water rice. *Plant Physiol.*, 72: 441–446.

Minami, A., Yano, K., Gamuyao, R., Nagai, K., Kuroha, T., Ayano, M. et al. (2018). Time-course transcriptomics analysis reveals key responses of submerged deepwater rice to flooding. *Plant Physiol.*, 176(4): 3081–3102.

Miro, B., and Ismail, A.M. (2013). Tolerance of anaerobic conditions caused by flooding during germination and early growth in rice (*Oryza sativa L.*). *Front. Plant Sci.*, 4: 269.

Munda, Sushmita, Nayak, A.K., Mishra, P.N., Bhattacharyya, P., Mohanty, Sangita, Kumar, Anjani et al. (2016). Combined application of rice husk biochar and fly ash improved the yield of lowland rice. *Soil Research*. http://dx.doi.org/10.1071/SR15295.

Narsai, R., Wang, C., Chen, J., Wu, J., Shou, H., and Whelan, J. (2013). Antagonistic, overlapping and distinct responses to biotic stress in rice (*Oryza sativa*) and interactions with abiotic stress. *BMC Genomics*, 14: 93. doi: 10.1186/ 1471-2164-14-93.

Neeraja, C.N., Maghirang-Rodriguez, R., Pamplona, A., Heuer, S., Collard, B.C.Y., Septiningsih, E.M., Vergara, G. et al. (2007). A marker-assisted backcross approach for developing submergence-tolerant rice cultivars. *Theor. Appl. Genet.*, 115: 767–776.

Paparella, S., Arau, J.S.S., Rossi, G., Wijayasinghe, M., Carbonera, D., Balestrazzi, A. (2015). Seed priming: state of the art and new perspectives. *Plant Cell Rep.*, 34: 1281–1293.

Pearce, D.M.E., and Jackson, M.B. (1997). Comparison of growth responses of barnyard grass (*Echinochloa oryzoides*) and rice (*Oryza sativa*) to submergence, ethylene, carbon dioxide and oxygen shortage. *Ann. Bot.* 68: 201–209.

Pearce, B.C., Parker, R.A., Deason, M.E., Qureshi, A.A. and Wright, J.J. K. (1992). Hypocholesterolemic activity of synthetic and natural tocotrienols. *J. Med. Chem.*, 35: 3595–3606.

Pucciariello, C., Voesenek, L.A., Perata, P., and Sasidharan, R. (2014). Plant responses to flooding. *Front. Plant Sci.*, 5: 226.

Raskin, I., and Kende, H. (1984). Role of gibberellin in the growth response of submerged deep-water rice. *Plant Physiol.*, 76: 947–950.

Sarkar, R.K., Reddy, J.N., Sharma, S., and Ismail, A.M. (2006). Physiological basis of submergence tolerance in rice and implications for crop improvement, *Curr. Sci.*, 91(7): 899–906.

Sarkar, R.K., Panda, D., Reddy, J.N., Patnaik, S.S.C., Mackill, D.J., and Ismail, A.M. (2009). Performance of submergence tolerant rice genotypes carrying the Sub1 QTL under stressed and non-stressed natural field conditions. *Indian Journal of Agricultural Science*, 79: 876–883.

Sarkar, R.K., and Bhattacharjee, B. (2011). Rice genotypes with *SUB1 QTL* differ in submergence tolerance, elongation ability during submergence and re-generation growth at re-emergence. *Rice*, 5(1): 7.

Schaller, G.E., and Bleecker, A.B. (1995). Ethylene-binding sites generated in yeast expressing the Arabidopsis ETRl gene. *Science*, 270: 1809–1811.

Septiningsih, E.M., Pamplona, A.M., Sanchez, D.L., Neeraja, C.N., Vergara, G.V., Heuer, S. et al. (2009). Development of submergence-tolerant rice cultivars: The Sub1 locus and beyond. *Annals of Botany*, 103: 151–160.

Septiningsih, E.M., Collard, B.C.Y., Heuer, S., Bailey-Serres, J., Ismail, A.M., and Mackill, D.J. (2013). Applying genomics tools for breeding submergence tolerance in rice. pp. 9–30. *In:* Varshney, R.K., Tuberosa, R. (Eds.). *Translational Genomics for Crop Breeding: Abiotic Stress, Yield and Quality*, 1st ed., Chichester, UK: John Wiley and Sons, Inc..

Setter, T.L., Ella, E.S., and Valdez, A.P. (1994). Relationship between coleoptile elongation and alcoholic fermentation in rice exposed to anoxia. II. Cultivar differences. *Annals of Botany*, 74: 273– 279.

Setter, T.L., Waters, I., Sharma, S.K., Singh, K.N., Kulshreshtha, N., Yaduvanshi, N.P.S. et al. (2009). Review of wheat improvement for waterlogging tolerance in Australia and India: The importance of anaerobiosis and element toxicities associated with different soils. *Annals of Botany*, 103: 221–235.

Shahzad, Z., Canut, M., Tournaire-Roux, C., Martinière, A., Boursiac, Y., Loudet, O. et al. (2016). Potassium-dependent oxygen sensing pathway regulates plant root hydraulics, *Cell*, 167(1): 87–98.e14.

Siangliw, M., Toojinda, T., Tragoonrung, S., and Vanavichit, A. (2003). Thai Jasmine rice carrying QTLCh9 (Sub1 QTL) is submergence tolerant. *Annals of Botany*, 91: 225–261.

Singh, S., Mackill, D.J., and Ismail, A.M. (2009). Responses of Sub1 rice introgression lines to submergence in the field: Yield and grain quality. *Field Crops Res.*, 113: 12–23.

Singh, S., Mackill, D.J., and Ismail, A.M. (2011). Tolerance of long-term partial stagnant flooding is independent of the Sub1 locus in rice. *Field Crops Res.*, 121: 311–323.

Singh, B.N., Dhua, S.R., Sahu, R.K., Patra, B.C. and Marndi, B.C. (2001). Status of rice germplasm – Its collection and conservation in India, *Indian J. Plant. Genet. Resour.*, 14: 105–106.

Singh, A.K., Gopalakrishnan, Singh, V.P., Prabhu, K.V., Mahapatra, T., Singh, N.K. et al. (2011). Marker assisted selection: A paradigm shift in basmati breeding. *Indian J. Genet. Plant Breed*, 71(2): 120.

Singh, Kuldeep, Mangat, G.S., Kaur, Rupinder, Vikal, Yogesh, Bhatia, Dharminder, Singh, Naveen et al. (2014a). Punjab Basmati 3: A bacterial Blight resistant dwarf version of basmati rice variety Basmati 386. *J. Res. Punjab Agric. Univ.*, 51: 206–207.

Singh, N., Khanna, R., Kaur, R., Lore, J.S., Neelam, K., Rani, N.S. et al. (2014b). Evaluation of aromatic rice germplasm against emerging pathotypes of Xanthomonas oryza pv.oryzae causing bacterial blight in Punjab. *In: Proceedings of National Symposium on Crop Improvement for Inclusive Sustainable Development*, pp. 314–315.

Singh, N., Dang, T., Vergara, G., Pandey, D., Sanchez, D., Neeraja, C. et al. (2010). Molecular marker survey and expression analysis of the rice submergence tolerance genes SUB1A and SUB1C. *Theor. Appl. Genet.* 121: 1441–1453.

Singh, A., Septiningsih, E.M., Balyan, H.S., Singh, N.K., and Rai, P.V. (2017). Genetics, physiological mechanisms and breeding of flood-tolerant rice (*Oryza sativa* L.). *Plant and Cell Physiology*, 58: 185–197.

Sripongpangkul, K., Posa, G.B.T., Senadhira, D.W., Brar, D., Huang, N., Khush, G.S. et al. (2000). Genes/QTL affecting flood tolerance in rice. *Theor. Appl. Genet.*, 101: 1074–1081.

Sultana, T., Ahamed, K.U., Naher, N., Islam, M.S., and Jaman, M.S. (2018). Growth and yield response of some rice genotype under different duration of complete submergence. *J. Agric. Eco. Res. Int.*, 15: 1–11.

Tamang, B.G., and Fukao, T. (2015). Plant adaptation to multiple stresses during submergence and following de-submergence. *Int. J. Mol. Sci.*, 16: 30164–30180.

Thomson, M.J., de Ocampo, M., Egdane, J., Rahman, M.A., Sajise, A.G., Adorada, D.L. et al. (2010). Characterizing the Saltol quantitative trait locus for salinity tolerance in rice. *Rice*, 3: 148–160.

Turner, N. (1981). Techniques and experimental approaches for the measurement of plant water status. *Plant and Soil*, 58: 339–366.

Vartapetian, B.B., and Jackson, M.B. (1991). Plant adaptation to anaerobic stress. *Ann. Bot.*, 79(suppl.A): 3–20.

Vergara, G.V., Mackill, D.J., and Ismail, A.M. (2014). Variation in tolerance of partial stagnant flooding in rice. *AoB PLANTS*, 6: Plu055.

Voesenek, L.A.C.J., and Bailey-Serres, J. (2013). Flooding tolerance: O_2 sensing and survival strategies. *Current Opinion in Plant Biology*, 16: 1–7.

Voesenek, L.A.C.J., and Bailey-Serres, J. (2015). Flood adaptive traits and processes: An overview. *New Phytologist*, 206: 57–73.

Wang, M., Lu, Y., Botella, J.R., Mao, Y., Hua, K., and Zhu, J.K. (2017). Gene targeting by homology-directed repair in rice using a Geminivirus-based CRISPR/Cas9 System. *Mol. Plant*, 10(7): 1007–1010.

Winkel, A., Colmer, T.D., Ismail, A.M., and Pedersen, O. (2013). Internal aeration of paddy field rice (*Oryza sativa*) during complete submergence—importance of light and floodwater O_2. *New Phytol.*, 197: 1193–1203.

Xu, K., Xu, X., Fukao, T., Canlas, P., Maghirang-Rodriguez, R., Heuer, S., Ismail, A.M. et al. (2006). Sub1A is an ethylene-response-factor-like gene that confers submergence tolerance to rice. *Nature*, 442(7103): 705–8.

Yamauchi, M., Aguilar, A.M., Vaughan, D.A., and Seshu, D.V. (1993). Rice (*Oryza sativa* L.) germplasm suitable for direct sowing under flooded soil surface. *Euphytica*, 67: 177–184.

Yang, C., Li, W., Cao, J., Meng, F.,Yu, Y., Huang, J. et al. (2017). Activation of ethylene signaling pathways enhances disease resistance by regulating ROS and phytoalexin production in rice. *Plant J.*, 89: 338–355.

Zheng, Y., Crawford, G.W., Jiang, L., and Chen, X. (2016). Rice domestication revealed by reduced shattering of archaeological rice from the lower yangtze valley. *Scientific Reports*, 6: 28136. Doi: 10.1038/srep28136.

4

Adaptive Morphology

Due to heterogeneity of flood-prone ecosystem in tropical Asian countries, like India, few indigenous rice landraces are still maintained and cultivated by poor farmers (Ram et al., 2002; Barik et al., 2020). Although such farming has poor yield capacity, this local landrace has excellent adaptation to extreme water availability, tolerant to different kinds of flooding and very important for quantitative trait loci (QTL) mapping and gene discovery (Singh et al., 2017).

Flooding causes many complex abiotic stresses (Jackson and Ram, 2003; Sarkar et al., 2006; Bailey-Serres et al., 2010), and the amount of harm that may be caused to the inundated plants varies, depending on flood water characteristics, like temperature, turbidity, water depth, oxygen, carbon dioxide concentration and light intensity. Temperature is another important factor which plays a vital role in rice plant's survival under flooding. The low temperatures (20°C) improve plant survival whereas the high temperature (30°C) increases plant mortality. The solubility of O_2 and CO_2 in flood water decreases at high temperatures. High temperatures speed up anaerobic respiration, resulting in starvation and death of the plants in a short duration (Ram et al., 2002).

Morphological

- Shoot elongation
- Root growth
- Leaf extension
- Aerenchyma formation
- Decrease in biomass

4.1 Seed Germination

Submergence of water at rice germination stage is a common practice in rice cultivation. Rice with flood tolerance at the germination stage, referred to as anaerobic germination (AG) tolerance, shows rapid coleoptile elongation under complete submergence with concomitantly delayed radicle development, enabling transportation of oxygen from the tip of the coleoptile above the water surface to the developing embryo. This allows rice to germinate under water and elongate the coleoptile to remain above the water surface to avoid anoxia.

Although seeds of the vast majority of higher plant species fail to germinate under anaerobic conditions, rice germinates successfully even when deprived of oxygen completely (anaerobiosis) to create the metabolic state of anoxia. According to Raymond et al. (1985), seeds can be grouped into classes on the basis of their responses to oxygen availability. Starchy seeds were shown to be especially tolerant of anaerobiosis because they are able to maintain a high energy metabolism under oxygen deficiency when compared with fatty seeds. However, amongst starchy seeds, there is considerable variation in the ability to germinate when anoxic. While declining oxygen concentrations negatively affect oat and barley germination (which starts with root emergence), rice behaves differently – root growth is suppressed while shoot growth increases as the oxygen concentration declines (Tsuji, 1973; Alpi and Beevers, 1983). Indeed, when germinated without oxygen, the final length of rice coleoptiles can exceed the length of aerobic coleoptiles whilst root and primary leaf fail to grow (Alpi and Beevers, 1983). This phenomenon is thought to increase the probability of the hollow coleoptile making contact with better-aerated water surface (the Snorkel effect; Kordan, 1974) thereby allowing oxygen to diffuse internally to the root and endosperm (Turner et al., 1981) and supporting more complete and vigorous seedling establishment (Fig. 4.1).

Submergence during germination is the first category of floods, which is often referred to as anaerobic germination (AG). This condition commonly occurs in rain-fed lowland ecosystems, which are characterized by heavy and sudden rainfall mostly associated after direct seeding cultural practices. Under this condition, rice seeds are submerged completely and suffer anoxia or hypoxia, which results in low or poor germination, poor crop establishment, and seedling death (Ismail et al., 2008). However, Angaji et al. (2010) reported rice germplasm with the ability to germinate under anoxia and hypoxia conditions. Complete submergence during the vegetative growth stage is the second category which occurs either at seedling or tillering stage. Plants are completely flooded, ranging from days to weeks before the flooding recedes. Singh et al. (2017) defined flooding tolerance in this case as the plant's ability to survive 10 to 14 days of complete submergence and continue their growth process after the flood recedes, with little damage to plant morphology. This problem is endemic in river basin areas, and it often leads to complete yield loss due to flash flood. Xu et al. (2006) cloned the Sub1 gene from the FR13A landrace from India, which was adapted to submerge at the vegetative

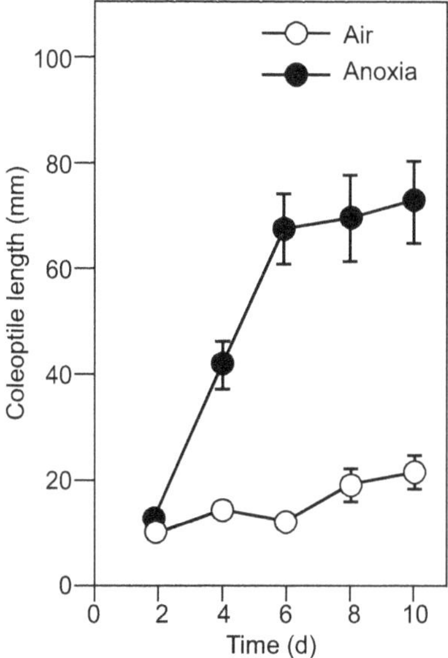

Fig. 4.1: Changes in coleoptile length (mm) with time.

stage. The third category of flooding occurs in lowland rice cultivation, where the water level ranges from 20 to 50 cm depending on whether it is a rain-fed or irrigated system. However, due to seasonal changes and rain duration, the water level might exceed 50 cm, up to 4m, which lasts for a prolonged period of time and could partially or completely submerge the plant, depending on the growth stage. This type of flooding may remain for few months depending on the area. Many landraces, also known as floating rice, are adapted to this type of flooding and can extend up to 25 cm per day as the flood level increases (Singh et al., 2017).

Among these three phases, the second phase is most critical which initiate at about 12 hours post imbibition of water and culminates at about 48 hours post imbibition of water. During this phase, the embryo prepares itself for germination and the food reserve stored inside the seed is transformed to a state that can support the germination process (He and Yang, 2013). In aerobic condition oxygen uptake also follows similar trend as that of water uptake (Narsai et al., 2013). This absorbed oxygen is used for aerobic respiration which yields energy required for seed germination. But when the oxygen supply is restricted, the germination process falls back to anaerobic respiration as an alternative resort to fulfil energy demand (Narsai et al., 2013). So, the success of GSOD tolerance largely depends on the flexibility of reprogramming respiration process from aerobic to anaerobic mode.

Generally, stagnant flooding occurs as a result of heavy rain, which can be deep-water flooding or flash flooding. Under deep-water flooding, plants increase their stem length (internode) to escape the submergence. However, after the flood subsides, affected plants are prone to lodging and eventually this leads to death due to the exhaustion of reserved nutrients within a few days. Contrarily, during flash flood, plant growth is restricted and resumes after occurrence of the flood (Winkel et al., 2013).

In direct seeding of rice, poor seedling establishment is a major constraint because the germinating seeds are subjected to various stresses, such as hypoxia caused by flooding or submergence (Yamauchi et al., 1993; Ismail et al., 2009). Therefore, the morpho-physiological traits at the germination and post-germination stages, including the germination speed, seedling emergence rate, and development rate of coleoptiles and leaves, are important factors in successful seedling establishment under hypoxic conditions (Ogiwara and Terashima, 2001; Mori et al., 2012).

It was investigated the morpho-physiological traits of rice (*Oryza sativa* L.) during the germination and post-germination phases to explore avoidance of hypoxic conditions. Some researchers compared four lines selected for anaerobic germination (AG lines) with the variety IR42. The germination capacity of AG lines was higher than that of IR42. The germination percentages and coleoptile elongation differed among the four AG lines; IR06F459 showed the fastest germination and rapid coleoptile elongation. The coleoptiles of IR06F459 were significantly longer than those of IR42. The α-amylase activity in germinating seeds was significantly higher in IR06F459 than in IR42. At two days after sowing, the sucrose and glucose concentrations in germinating seeds were higher in IR06F459 than in IR42. These results show that IR06F459, an AG line with a long coleoptile, has high α-amylase activity and high sucrose and glucose concentrations in germinating seeds (Fig. 4.2). These attributes partly explain its vigorous germination and coleoptile growth under hypoxic conditions.

The semi-aquatic nature of rice plant makes it capable of growing under waterlogged and/or submerged condition for a considerable period. This is possible due to elongation of submerged shoot organs at faster rates by developing aerenchyma, that allows sufficient internal transport of oxygen to submerged plant parts from the re-emerged elongated shoot (Jackson and Ram, 2003; Magneschi and Perata, 2009). The process of energy generation and detoxification of fermented products through coleoptile elongation leads to survival under the conditions of low oxygen (Miro et al., 2017). The germination and early seedling growth stage of rice have been found to be highly intolerant to submergence (Ismail et al., 2009; Angaji et al., 2010; Joshi et al., 2013). Apart from germination, submergence also brings many morphological and physiological changes in rice plant. During submergence, rice plant survives by elongation of leaf sheath and blade at seedling stage and internodes at vegetative growth stage (Haque, 1974). If the flooding duration exceeds two to three

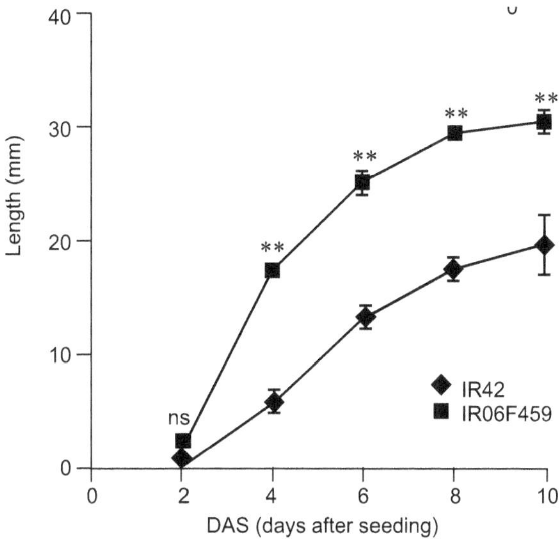

Fig. 4.2: Coleoptile length of IR06F459 is higher than IR42 under submergence.

weeks, even the submergence-tolerant varieties try to expose their leaf tip above the water surface for ensuring survival (Sarkar and Bhattacharjee, 2011; Colmer et al., 2014).

Soil oxygen deficiency has indirect effects on plants by promoting the growth of anaerobic bacteria. Some anaerobes reduce ferric ion (Fe^{3+}) to ferrous ion (Fe^{2+}), and due to greater solubility of Fe^{2+}, Fe^{2+} concentrations can rise to toxic level. Other anaerobic microorganisms may reduce sulphate (SO_4^{2-}) to hydrogen sulphide (H_2S), which is a respiratory poison. These toxic substances cause severe damage to plants by reducing the redox potential under flooding conditions and increase the susceptibility to diseases (Ram et al., 2002). Also, the acetic acids as well as butyric acids released by anaerobic microorganisms are toxic to plants at high concentrations.

4.2 Root Growth

Flooding negatively affects rice production in over 20 million hectares of rain-fed lowlands and flood-prone areas in Asia. While there are numerous reports on the response of rice shoots to flooding, scant information is available on the effect of flooding on rice roots. This study assessed the effect of complete submergence at the vegetative stage on growth and physiological responses of rice roots. Seedlings of four rice genotypes—tolerant varieties FR13A and Swarna-Sub1 and sensitive varieties Swarna and IR42—were completely submerged in a concrete tank for 12 days. Afterwards, water was drained and seedlings were allowed to recover. Survival was recorded 14 days after de-submergence. Seedlings were considered surviving when they are able to generate new leaves. Root measurements conducted during submergence were: elongation, root viability, peroxidase activity, and membrane damage assessed as concentrations of malondialdehyde and electrolyte leakage. Seedlings of tolerant genotypes had higher survival, and the roots were more viable with greater capacity to elongate. Moreover, they had higher peroxidase activity and lesser increases in both electrolyte leakage and malondialdehyde production during submergence. There were strong positive correlations between survival and some parameters measured during submergence, such as root elongation (r = 0.74**) and peroxidase activity (r = 0.79**, at day seven of submergence). Strong negative correlations were observed between survival and membrane damage during submergence (r = –0.83** and –0.79** for malondialdehyde levels and electrolyte leakage, respectively). Data showed that tolerance to complete submergence at the vegetative stage may be associated with some root traits, such as high viability and high activity, translated into greater elongation growth and lesser membrane damage during submergence. Such traits might play important roles in maintaining root function in rice seedlings exposed to complete submergence during the vegetative stage (Fig. 4.3).

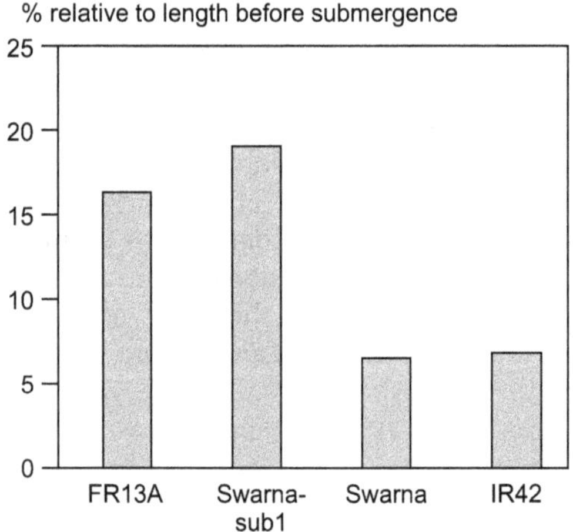

Fig. 4.3: Root elongation of rice varieties during submergence.

4.3 Seedling Growth

Numerous studies have examined the influence of flooding on rice tillering (Huo et al., 1997; Sasidharan and Voes enek, 2013; Wang et al., 2014). Rice plants at the tillering stage possess sufficient carbohydrates for growth; compared with controls, plants in the late tillering stage treated with two days of submergence did not show a significant difference in the indices investigated, except for an increase in the vitality of the root system. Rice plants at the tillering stage have higher submergence tolerance than those at other stages, and submergence at this stage has little influence on rice yield. Thus, it may be possible to deepen post-rainfall water storage containers and expand the water content control space in paddy fields (Huo et al., 1997; Sasidharan and Voesenek, 2013; Wang et al., 2014). According to the literature, rice submergence can result in yellow leaf discoloration, a decrease in the number of green leaves and white roots, a decrease in root absorption, impaired growth and development, and a decrease in lodging tolerance, which can result in underproduction or even a complete lack of yield (Vriezen et al., 2003; Kawano et al., 2009; Colmer and Voesenek, 2009; Crawford, 2003).

In this study, 60 direct seeding rice varieties with different genotypes were used as experimental materials. Through long-term submergence treatment (seven days and 14 days), the responses of different genotypes to submergence stress were compared in germination and seedling emergence, stem, leaf and root elongation, starch and other storage material consumption. The results showed that there were significant differences in submergence tolerance of different genotypes at the germination and seedling stage. Through further analysis by statistical methods, such as correlation analysis, principal component analysis, weight comprehensive evaluation, and cluster analysis, 60 varieties could be divided into four categories: strong submergence tolerance, medium submergence tolerance, weak submergence tolerance and submergence sensitivity, while 6, 22, 4 and 28 varieties were screened respectively (Fig. 4.4). After 14 days and 4 cm deep submergence, more than 78.1% of rice seeds could germinate, and the average plant height and root length could reach more than 11.99 cm and 9.66 cm respectively. The dry matter mass, starch content and soluble sugar content per plant were significantly higher than those of other tolerant types. The average coleoptile of the sensitive type was only 3.17 cm, and the radicle had little elongation. The evaluation and variety screening of submergence tolerance of different genotypes of direct seeding rice at the seedling stage can provide a theoretical basis for further clarifying the mechanism of rice-submergence tolerance, screening suitable direct seeding varieties and cultivating special varieties of direct seeding rice (Table 4.1).

Bui et al. (2019) highlighted the importance of Sub1 QTL in regulation of root physiology of rice under flooding conditions and indicated that the tolerant varieties show higher root activities, like root tip viability and root peroxidase, which result in minimum damage to the roots as well as shoots under flooding.

Flooding severely restricts O_2 and CO_2 gas exchange between rice tissues and atmosphere, inhibits aerobic respiration as well as photosynthesis, and accelerates the consumption of energy reserves,

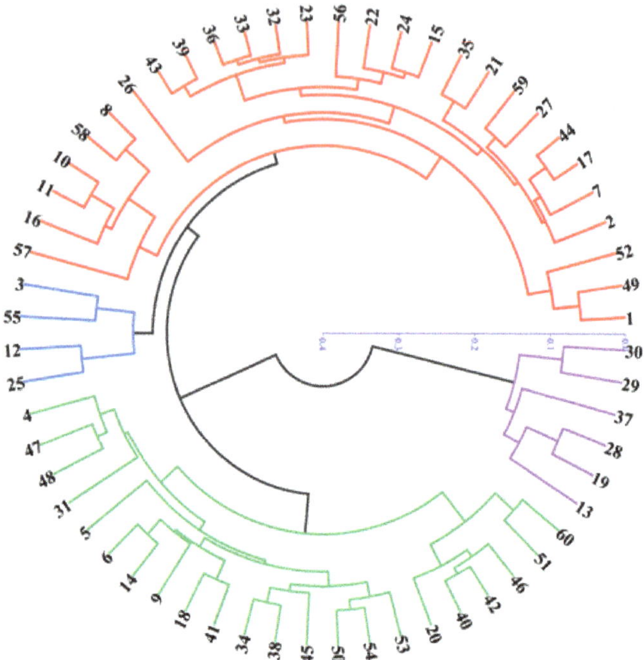

Fig. 4.5: Cluster diagram of submergence tolerance of direct seeding rice of different genotypes at seedling emergence stage. Violet-Rank category 1; Green-Rank category 2; Blue-Rank category 3; Red-Rank category 4.

leading to stunted growth and death (Kato et al., 2014). Flooding inhibits dry matter accumulation in susceptible genotypes (Singh et al., 2014). Reduction in growth is a common phenomenon when plants are flooded with water. However, tolerant rice genotypes have the potentiality to store sufficient amounts of dry matter and significantly produce more primary branches per panicle yield and harvest index under drought stress as compared to Nagina-22. The remaining Sub1 genotypes also showed better performance than drought-susceptible check (IR-64) and showed non-significant difference with Nagina-22 for most of the drought-tolerance-related traits (Fig. 4.4). The results recommended that genotype having Sub1 genes can effectively be grown under rain-fed regions which are equally prone to floods and drought (Fig. 4.6).

Vergara and Mazaredo (1975) identified some high submergence-tolerant rice varieties from Sri Lanka after seven-day submergence of 10-day-old seedlings and measured the survival rate of recovery after five days. Highly tolerant varieties showed more than 75% survival rate. Some of these varieties, such as Thavalu, Kurkaruppan, and Goda Heenati, have been extensively used in crop improvement and studies of submergence tolerance. Through standardized methods of screening at the IRRI,

Table 4.1: Root characters in rice under different levels of submergence and duration (days)

Submergence depth-duration (d)	Aboveground dry weight (g/hole)	Adventitious roots (number/hole)	Dry root weight (g/hole)	Length of the longest root (cm)	Root width (mm/bar)	Relative expression of the OS-ACS3 gene
Control-1	10.42 bc	387 a	1.661 ab	23.5 ab	0.973 bc	1.7 h
Control-3	11.34 ab	385 a	1.727 ab	24.1 ab	0.967 bc	1.6 h
Control-5	12.82 a	388 a	1.704 ab	24.4 a	0.987 abc	1.6 h
1/2-1	11.21 ab	357 ab	1.792 a	24.0 ab	0.960 c	11.4 e
1/2-3	10.81 ab	348 ab	1.688 ab	23.2 abc	0.947 c	6.3 f
1/2-5	10.55 abc	339 a	1.587 ab	22.0 bcd	1.003 abc	3.6 gh
2/3-1	10.11 bc	320 b	1.598 ab	22.0 bcd	0.953 c	16.6 d
2/3-3	9.68 bcd	326 b	1.334 abc	21.3 cde	1.033 ab	10.0 e
2/3-5	8.30 cd	321 b	1.243 bc	20.2 de	1.047 a	5.9 fg
1/1-1	10.50 abc	399 a	1.685 ab	21.4 cde	0.970 bc	90.3 a
1/1-3	10.19 bc	336 ab	1.512 ab	19.9 de	1.010 abc	35.9 b
1/1-5	7.81 d	331 b	0.992 c	19.4 e	1.013 abc	23.3 c

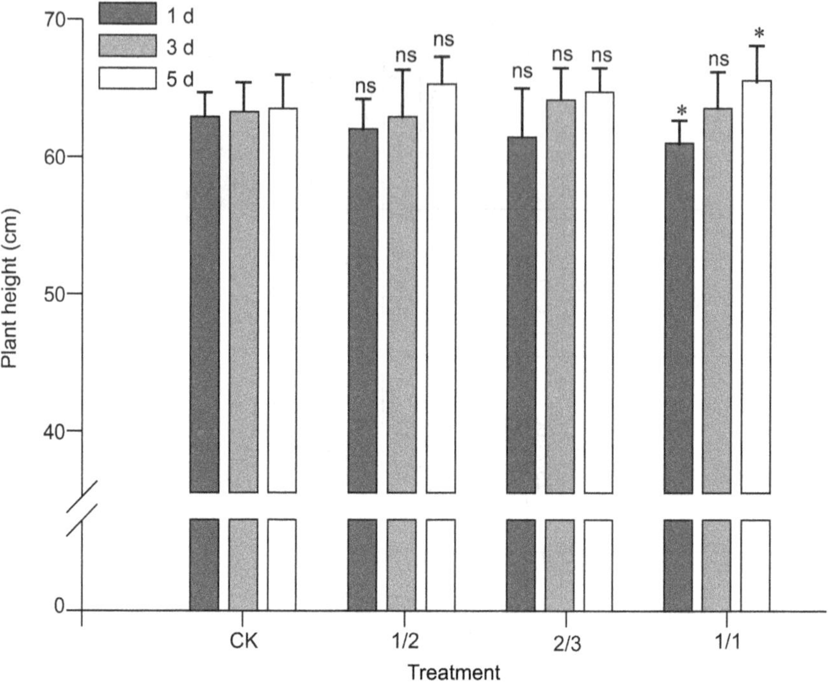

Fig. 4.6: Plant height of rice under varying levels of submergence and duration.

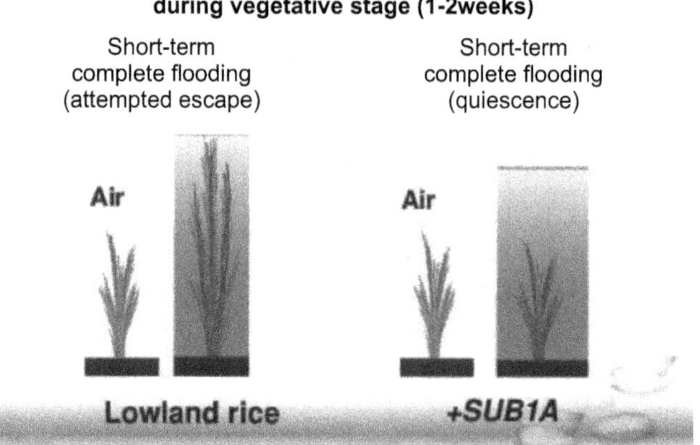

Fig. 4.7: Model showing attempted escape and quiescence in rice.

landraces tolerant to flooding at the germination stage were identified, which include Ma-Zhan Red, Nanhi, Khao Hlan On, and Khaiyan (Angaji et al., 2010). These rice germplasms have been used in physiological studies and genetic mapping for the development of improved rice lines with high levels of tolerance to aerobic germination stress, using molecular and conventional breeding methods (Toledo et al., 2015). In India, rice varieties, such as Sabita, Ambika, Saraswati, and Hangseswari are recommended by the Indian Council of Agricultural Research for deep-water rice farmers. These varieties, such as Hangseswari, can produce up to 2.5 t/ha grain yield under deep-waterlogged conditions (www.icar.org.in). Some varieties, such as Leuang Pratew 123 and Khao Tah Haeng 17, are capable of tolerating both stagnant floods of up to 80 cm and flash floods during the early vegetative growth stage. Although these varieties do not possess internode elongation properties, they are widely cultivated in low-lying areas. Some of the major advantages of these varieties contributing to flood tolerance include tolerance to drought, flexibility in planting time, tolerance to soil-related stressors, and plant height 150–180 cm to withstand 80 cm of water (Puckridge et al., 2000).

4.4 Tiller Growth

Two Indonesian rice cultivars were examined for growth characteristics under several durations of submergence consisting of 0 (control), 3, 5, 7, and 10 days. Rice seedlings were first submerged into a 45 cm water depth and then the water level remained constant, allowing some parts of the leaves to appear from the water surface during the treatments which resembled the actual flash-flooding conditions in the field. Results showed that the change in the relative growth rate (RGR) in both cultivars was attributed more to the net assimilation rate (NAR) alteration compared to the leaf area ratio (LAR). The relationship between NAR and specific leaf area (SLA) was also investigated, resulting in a significant negative correlation between these two variables where NAR increased by the thickening of the leaf blade. 1R Dadahup then was found to show a higher ability to survive and grow during a longer period of flash flooding, compared to Ciherang, based on the minimal sheath elongation under submergence and positive response in both SLA and the relative tillering rate (RTR) (Wati et al., 2016).

The main strategies enabling rice plants to cope up with flash-flooding stress require growth regulation during submergence and subsequent rapid growth recovery after de-submergence and better nutrient management options can enhance this strategy. Application methods of nitrogen and phosphorus were evaluated in submerged rice for tiller mortality, productivity, grain quality, and nutrient absorption. The performance of Sub1 (IR-64 Sub1 and Swarna Sub1) and non-Sub1 (IR-20) cultivars of rice was tested under clear and turbid water for their tolerance to submergence in response to basal phosphorus and post-submergence nitrogen. Tillering ability, yield, and nutrient absorption of rice subjected to complete submergence for 15 days decreased significantly over non-submerged rice plants. Turbid water submergence was fatal in terms of tiller mortality, reduced nutrient absorption, and yield because of low light and dissolved oxygen underwater. The crop fertilized with nitrogen produced more number of tillers, yield attributes and grain yield than the unfertilized crop.

Of the three growth stages, viz. seedling establishment, maximum tillering and flowering, complete submergence of plant at flowering stage was found to be most critical followed by seedling establishment and maximum tillering stages. Among the three stages of reproductive growth phase, booting stage was found to be most susceptible to complete submergence followed by flowering and post-flowering. The submergence at booting for four days was equally detrimental as that of six or eight days at flowering. Irrespective of the growth stages, the plants subjected to complete submergence showed higher nitrogen content (in plant as well as in grain) as compared to those grown under control conditions (5 ± 2 cm) and increased with the increase in duration of submergence. The P and K contents in the plant decreased under submergence. Damage is most severe during flowering. When flooding occurs at flowering, fertilization would not occur resulting to total grain yield loss.

Recent climate change projections anticipate spatial shift in precipitation pattern and increase of flooding events that may have negative effects on rice yield and economic returns. Efforts for increasing rice production in areas prone to submergence stress will directly benefit hundreds of millions of people dependent on rice as their staple food. This necessitates an objective review of physiological mechanisms and management practices, which could sustain crop productivity under partial or complete submergence. Submergence usually reduces photosynthesis rate that results in quick depletion of the carbohydrate reserve and ultimately the plant dies. Varieties introgressed with *SUB1A* QTL maintain higher activity of alcohol dehydrogenase and low rate of chlorophyll degradation and thus exhibit better survival under submergence. Prolonged submergence results in a significant reduction in soil redox potential and the heavy influx of flood water promotes runoff, volatilization, and deep percolation which leads to loss of sizeable amount of nutrients and ultimately causes nutrient deficiency in soil. Thus, to ensure optimum yield, it is essential to alter the nutrient schedule when the plant is subjected to submergence stress. Agronomic management practices, like seed priming, higher seed rates, alteration in crop geometry and other improved seeding methods, enhance production efficiency by boosting germination, early growth, and optimum partitioning of photosynthates to vegetative and reproductive parts.

4.5 Aerenchyma Formation

The main adaptation of lowland rice to soil waterlogging is the formation of aerenchyma, which permits relatively unhindered transport of O_2 from well-aerated shoots to submerged roots (Figs. 4.1, 4.2; Armstrong, 1979; Jackson and Armstrong, 1999). Longitudinal diffusion of O_2 towards the root apex can be further enhanced by induction of a barrier to radial O_2 loss (ROL) that minimizes loss of O_2 to the surrounding environment. Furthermore, this barrier may impede the movement of soil-derived toxins (i.e., reduced metal ions) and gases (e.g., methane, CO_2, and ethylene) into the roots (Armstrong, 1979; Colmer, 2003a; Greenway et al., 2006). Both upland and lowland rice species use these traits under waterlogged conditions (Colmer, 2003b).

In response to waterlogging, rice and many wetland plants form gas spaces called 'aerenchyma' by inducing the death of cells inside the roots, allowing oxygen to be transported from leaves down to the roots. Aerenchyma thus plays an important role in the survival of rice and other plants under waterlogged conditions.

Effective internal aeration in plants is crucial to survival under submergence. In rice, aerenchyma is well developed in roots, internodes, sheaths, and the mid-rib of leaves (Colmer and Pedersen, 2008a; Matsukura et al., 2000; Steffens et al., 2011) and contributes to the effective internal aeration between shoots and roots (Colmer, 2003a; Colmer and Pedersen, 2008a). Submerged leaves have gas films that aid O_2 and CO_2 exchange between leaves and the surrounding water, and thus increase underwater net photosynthesis by supplying CO_2 during the day (under light conditions) and promote O_2 uptake for respiration at night (Colmer and Pedersen, 2008b; Pedersen et al., 2009; Raskin and Kende, 1983). As a result, leaf gas films contribute to leaf sugar production by photosynthesis when under water, and in turn, shoot and root dry mass (Pedersen et al., 2009). Removal of leaf gas films caused a decrease of O_2 partial pressure (pO_2) in roots when shoots were in darkness, suggesting that leaf gas films partially contribute to O_2 transport from shoot to root (Pedersen et al., 2009; Winkel et al., 2011). Taken together, these findings indicate that leaf gas films are important for submergence tolerance (Pedersen et al., 2009; Raskin and Kende, 1983) In roots, internodes, sheaths, and the mid-rib of leaves (Colmer and Pedersen, 2008a; Matsukura et al., 2000; Steffens et al., 2011) and contributes to the effective internal aeration between shoots and roots (Colmer, 2003a; Colmer and Pedersen, 2008a). Submerged leaves have gas films that aid O_2 and CO_2 exchange between leaves and the surrounding water, and thus increase underwater net photosynthesis by supplying CO_2 during the day (under light conditions) and promote O_2 uptake for respiration at night (Colmer and Pedersen, 2008b; Pedersen et al., 2009; Raskin and Kende, 1983). As a result, leaf gas films contribute to leaf sugar production by photosynthesis when under water and in turn, shoot and root dry mass (Pedersen et al., 2009). Removal of leaf gas films caused a decrease of O_2 partial pressure (pO_2) in roots when shoots were in darkness, suggesting that leaf gas films partially contribute to O_2 transport from shoot to root (Pedersen et al., 2009; Winkel et al., 2011). Taken together, these findings indicate that leaf gas films are important for submergence tolerance in rice (Pedersen et al., 2009; Raskin and Kende, 1983) (Fig. 4.8).

Ethylene-responsive lysigenous aerenchyma formation is affected by chemical inhibitors or stimulators of programmed cell death (PCD) and other pathways. Curiously, only some cortical cells die while others do not during aerenchyma formation.

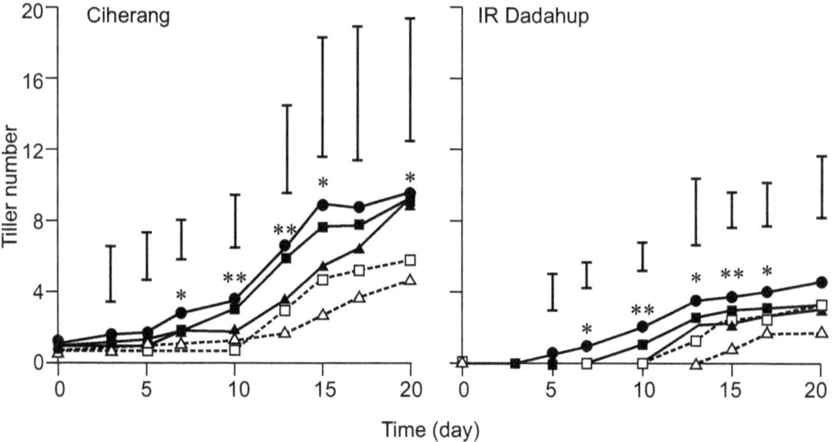

Fig. 4.8: 1R Dadahup showed greater ability to survive and grow during a longer period of flash flooding as compared to Ciherang.

4.6 Leaf Chlorophyll Content

An experiment was conducted in 2014 and 2015 *kharif* season to evaluate the performance of rice genotypes under submerged condition on the basis of morpho-physiological traits and yield attributes. The study revealed that rice variety, Swarna Sub 1, exhibited the highest survival (91.2%), whereas IR42 as susceptible check showed the least survival (4.2%) under submergence. Among the tested genotypes, IR 10F365, IR 11F216 and 11F239 with the respective survival values of 81.4, 80.0 and

78.1%, respectively, were found to be at par with Swarna Sub 1 after 16 days of genotype in complete submergence at vegetative stage. Moreover, physiological traits, like chlorophyll, sugar content and anti-oxidative system (SOD and CAT), were higher in tolerant genotypes as compared to susceptible ones. Less reduction in the content of sugar, chlorophyll, SOD, and CAT activity was observed in genotypes IR 10F365, IR 11F216 and IR 11F239 along with var. Swarna Sub 1 (tolerant check) after submergence, while in susceptible genotypes IR 09L311, IR 08L216 and IR55423-01 the reduction in sugar and chlorophyll content was higher just after submergence. Apart from physiological traits, tolerance genotypes have higher yield and yield-attributing character as compared to susceptible genotypes (Figs. 4.9 and 4.10) (Dwivedi et al., 2017).

An experiment was conducted with four rice genotypes: Swarna Sub1, Nagina-22, NDR-102 and NDR-97 and submergence-related traits were recorded at the student's instructional farm of Narendra Deva University of Agriculture and Technology, Kumarganj, Ayodhya (U.P.). Thirty-five-day-old direct seeded rice seedlings was exposed to 11 days of complete submergence treatment. Shoot elongation rate, catalase, and peroxidase activity was recorded at the end of submergence treatment. Less shoot elongation and high catalase activity were recorded in Swarna Sub1 and NDR-102 while Nagina-22 and NDR-97 showed high reduction and less increase respectively. Swarna Sub1and NDR-102 showed fast recovery after de-submergence. NDR-97 poorly survived and Nagina-22 failed to survive during submergence period. 50% flowering was less affected in Swarna Sub1 and NDR-102; it was delayed in the case of NDR-97. Grain yield per plant was highly affected in NDR-97 while less in Swarna Sub1 and NDR-102 (Fig. 4.10).

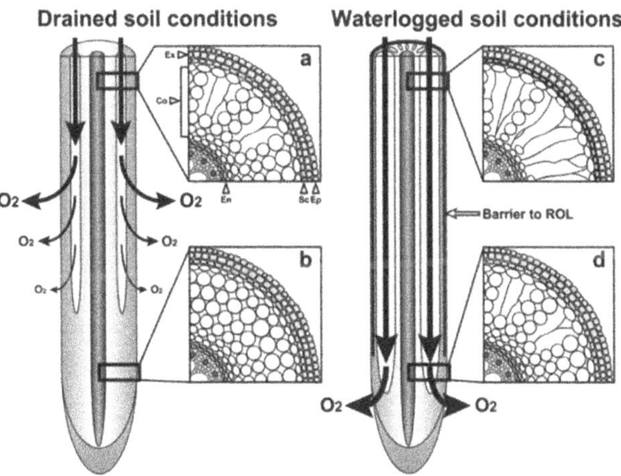

Fig. 4.9: Differences in lysigenous aerenchyma formation and patterns of radial O_2 loss (ROL) in rice roots under drained soil conditions and waterlogged soil conditions.

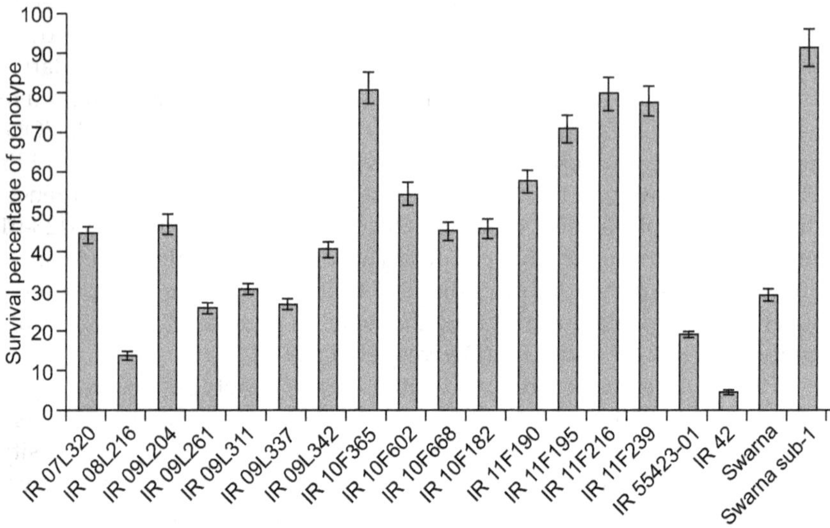

Fig. 4.10: Survival percentage of rice genotypes under 16 days of complete submergence.

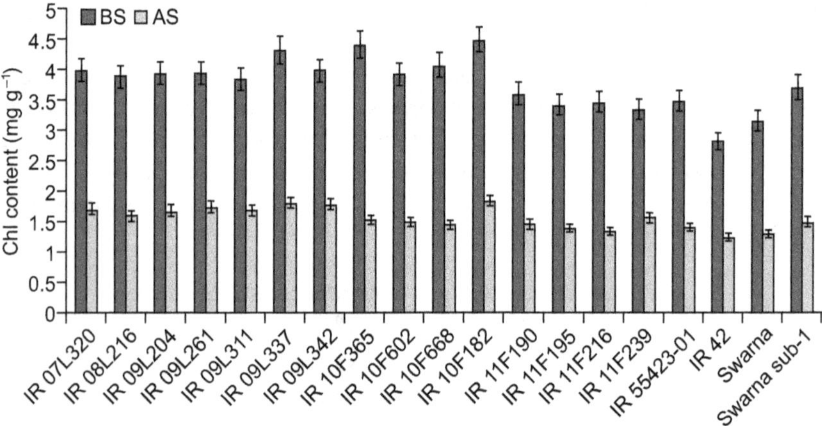

Fig. 4.11: Chlorophyll content of rice leaves before submergence (BS) and after submergence (AS).

An experiment was conducted during the period from July to December, 2017 in transplanting 'aman' season. The experiment was conducted in the Sher-e-Bangla Agricultural University Farm, Dhaka, Bangladesh to observe the effect of submergence at early vegetative stage on growth and yield of some transplanting 'aman' rice varieties. Four submergence duration, such as control (no submergence), six days' submergence, ten days' submergence and fourteen days' submergence and six transplanting 'aman' varieties, such as BRRI dhan51, BRRI dhan46, BRRI dhan34. BRRI hybrid dhan1, BRRI hybrid dhan2, ACI hybrid1 were

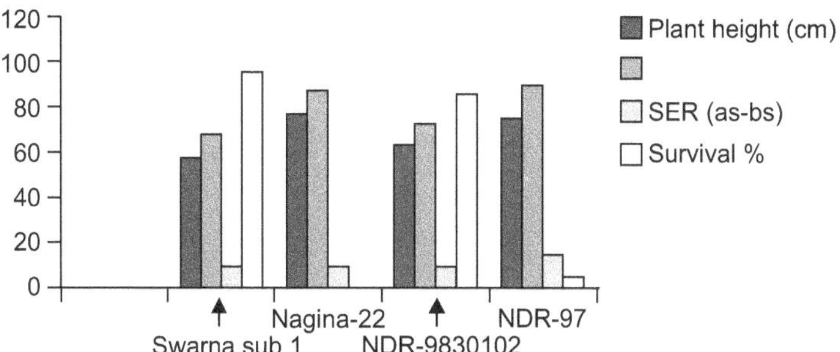

Fig. 4.12: Effect of submergence on shoot elongation (cm) and survival (%) in rice genotypes.

used in this experiment. The plant was submerged completely at early vegetative stage (30 days after transplanting) in unit plot to a depth of 40 cm above the soil level. The water level was higher than the plant height. The growth parameter leaves per plant showed reduction of number in each date of sampling with the duration of submergence (Fig. 4.12).

Based on the results of the present study, it was concluded that both variety and submergence duration had significant influence on morphological characters of rice at early vegetative stage. The tested genotypes showed wide variation in yield reduction per cent in different submergence duration. BRRI dhan51 followed by BRRI hybrid dhan1 showed lower grain yield reduction % in submerged conditions compared to control by attaining good yield contributing characters and thus proved as tolerant varieties. On the other hand, BRRI dhan34 and ACI hybrid1 were susceptible to submergence.

Global climate change has an impact on the frequency and magnitude of hydrological fluctuations, which can result in catastrophic events, such as floods and droughts, among other things. Extremes in precipitation, both high and low, are rapidly limiting food, fiber, and forest production across the planet (Mohanty et al., 2013). As a result, improving rice's combined resistance to submergence and drought will significantly enhance rice yield while also preserving water resources and soil quality (Xiong et al., 2019). In order to do this, genotypes containing the submergence resistant gene (Sub1) were assessed in fields under total submergence and severe drought conditions for yield-related characteristics (Table 4.2).

4.7 Programme for Genomic Resource Utilization

Floods have now become the most detrimental natural catastrophe worldwide due to radical climatic fluxes. Therefore, there is a dire necessity to develop high-yielding rice lines to deal with this scenario. For this purpose, a large-scale experiment was conducted, including 115 rice genotypes having *Sub*1 gene imported from International Rice

Table 4.2: Two-sided Dunnett's multiple comparisons for yield and yield related traits under drought.

Genotype/Parameter	Nagina22	Swarna sub1	IR44 sub1	IR-07-F289 sub1	Ciehrang sub1	IR64
Chlorophyll content	33.24	34	34.18	34.08	34.11	39.06
Leaf area(m^2)	3564.7	3867.7	2278.7*	4752.7*	2952.0	6142.7*
Plant height(cm)	76.66	50.65*	56.75*	58.31*	56.0*	81.66
Productive tillers per plant	13.66	10.5	18.83*	9.33	19.5*	8.66
Panicle length(cm)	15.42	14.5	19.21	18.26	19.2	25.42*
Primary brs./panicle	7.33	10.5*	8.66	8.83	7.33	8.83
Spikelets/panicle	87.33	97.17	69.67	78.00	89.00	157.30*
Spikelet fertility%	17.17	30.0	63.5	19.0	64.5	101.83
Biological yield/ plant(g)	106.57	39.63*	29.67*	35.83*	49.20*	112.93
Grain yield/plant(g)	3.17	5.13*	0.98*	3.93	2.31	1.07*
Harvest Index	3.02	13.09*	3.48	13.94*	4.85	0.96

Research Institute (IRRI), Philippines, six local cultivars/approved varieties and three high-yielding rice varieties, i.e., Sabitri, IR6 and NSICRC222 being used as potential varieties in different countries of Asia as susceptible check and IR64-*Sub*1 as tolerant check. The genotypic screening was performed using two PCR-based DNA markers, i.e., ART5 and SC3. Phenotypic screening was conducted in a natural pond to assess the interaction of *Sub*1 gene in natural stagnant flood water as well as the suitability of introgression of *Sub*1 gene in approved varieties and elite rice lines. The genotypes were assessed in terms of plant survival percentage, submergence tolerance index, physical condition, stem elongation, number of grains per panicle, thousand grain weight, grain yields, and deviations in these traits after submergence stress. The PCR results suggested that both the primers, ART5 and SC3, may be used as potential PCR-based markers for molecular screening of rice genotypes for *Sub*1 QTL. Furthermore, it confirmed the presence of *Sub*1 gene in all the lines imported from IRRI, while it was absent in all the local cultivars studied. All the genotypes with submergence-tolerant gene (*Sub*1) showed significantly greater tolerance level in submergence stress of 14 days, as compared to other local cultivars/varieties, authenticating the effectiveness of *Sub*1QTL in conferring submergence tolerance. Significantly different performances of all the *Sub*1 genotypes in terms of all the studied traits indicate high genotypic and Genotypic Environment Interaction (GEI) of *Sub*1QTL. Employment of *Sub*1 lines, such as R105479:149-18, IR64-

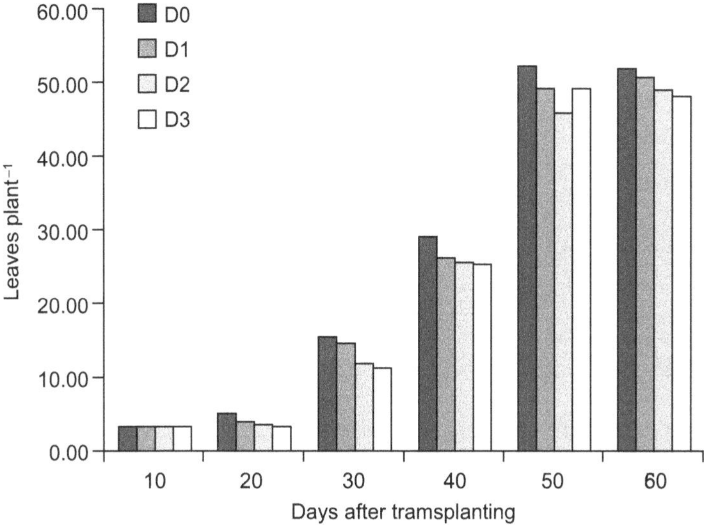

Fig. 4.13: Reduction of average leaves per plant under different duration of submergence.

*Sub*1 and RI05469:81-22-3 in breeding programs for developing flood-tolerant rice varieties might further upsurge rice yields in flash-flood areas. Correlation analysis revealed that plant survival percentage after submergence reduced stem elongation during submergence and submergence-tolerance index are very important traits for developing submergence-tolerant lines (Table 4.3 and Fig. 4.13).

To evaluate the genotypes by multiple agronomic parameters for submergence tolerance at different growth stages, 21 'aman' rice genotypes were tested against three submergence treatments, viz. (i) no submergence or control, (ii) submerge at 10 days after transplanting, (DAT), and (iii) submerge at 30 DAT. The experiment was laid out in a split plot design with three replications. The plants were completely submerged in the tanks for 14 days. Dry matter production, growth, yield, and yield components of rice were significantly reduced due to interaction effect of genotypes × submergence. Submergence at the early stage of growth (10 DAT) was more detrimental to plant growth than what happened at later stages (30 DAT). Wide genotypic differences in relative submergence tolerance based on grain and tiller numbers were identified. Grains and tiller numbers contributed most of the variations to seed yield among parameters investigated. When genotypes were ranked for submergence tolerance based on the means of multiple parameters, dramatic changes of submergence tolerance at early and later stages were observed in four genotypes; BRRI dhan33, Lalchikon, Achin and Sadamota were identified with a favourable combination of submergence tolerance (Table 4.4) (Abedin et al., 2019).

Table 4.3: Average (±S.E), minimum and maximum values of different traits of genotypes under normal and submerged conditions.

	Control			Submergence Stress		
Traits	**Average (± S.E)**	**Min.**	**Max.**	**Average (± S.E)**	**Min.**	**Max.**
Thousand Grain Weight (g)	23.24 ± 1.476	17.31	37.89	18.79 ± 1.264	2.98	36.28
Grains per panicle (g)	94 ± 5.621	51	171	40 ± 6.428	1	96
Potential Yield (t/ha)	4.95 ± 1.637	1.00	11.15	2.54 ± 1.370	0.089	8.00
Plant Height (cm)	105 ± 5.324	84	132	121 ± 4.925	100	151
Tillers per plant	15 ± 0.955	10	24	-	-	-

The present investigation to assess genetic diversity among 49 rice genotypes including four checks (FR13A, Hemavathi, Swarna Sub1 and Jyothi) was conducted at College of Agriculture, University of Agricultural and Horticultural Sciences (UAHS), Shivamogga during *kharif*, 2016. The clustering pattern based on Mahalanobis' D^2 analysis revealed that 49 genotypes were grouped into eight clusters where cluster VI was the largest cluster with 10 genotypes and cluster I was the smallest (1 genotype). The maximum intra cluster distance was exhibited by cluster VII (98.11). The maximum inter cluster distance was recorded between cluster VII and cluster VIII, suggesting that the genotypes constituted in these clusters can be used as parents for future hybridization programme, while the minimum was between cluster V and cluster VI (57.17). Cluster II showed highest mean values for the number of tillers per plant, number of productive tillers per plant, number of spikelets per panicle, number of filled grains per panicle, grain yield per plant, straw yield per plant. Traits like absolute growth rate, days to 50 per cent flowering, days to maturity, plant height, number of filled grains per panicle and grain yield per plant contributed 95.24 per cent towards the total genetic divergence among the genotypes. Hence these traits should be of foremost importance while selecting parents in hybridization programme (Tables 4.5 and 4.13) (Lahari et al., 2017).

Rice has served as a prime model plant species for studies of constitutive root aerenchyma. In an investigation, 20-day-old seedling root anatomy was studied in eight diverse and phenotypically contrasting rice genotypes. Studies were carried out in contrasting water regimes, i.e. aerobic and waterlogging situations. Higher number of aerenchyma formation was observed in waterlogging treatment as compared to aerobic condition. Thus, development of aerenchyma in response to waterlogged condition would help supply O_2 to root cells. Aerenchyma cells in aerobic roots ranged from 11 (BJ21) to 23 (Devamallige). However,

Table 4.4: Yield and yield attributes of rice varieties at different time of submergence.

Cultivars	Submergence at	Plant height	Tiller production	Shoot dry matter	Root dry matter	Crop growth rate	Relative growth rate
BRRI dhan33	10 DAT	1.22	0.45	0.49	0.33	0.96	0.30
	30 DAT	1.22	0.86	0.77	0.50	0.63	0.50
BRRI dhan51	10 DAT	1.29	0.30	0.37	0.23	0.74	0.11
	30 DAT	1.16	0.80	0.90	0.61	0.85	0.54
BRRI dhan56	10 DAT	1.09	0.39	0.35	0.28	0.64	0.14
	30 DAT	1.16	0.84	0.81	0.41	0.63	0.39
BRRI dhan57	10 DAT	1.12	.30	0.27	0.16	0.50	0.11
	30 DAT	1.08	0.66	0.61	0.28	0.43	0.27
BINA dhan11	10 DAT	1.20	0.30	0.37	0.22	0.59	0.11
	30 DAT	1.37	0.58	0.61	0.25	0.60	0.97
BU dhan1	10 DAT	1.55	0.48	0.40	0.36	0.65	0.10
	30 DAT	1.18	0.62	0.71	0.09	0.15	0.18
Guti shorna	10 DAT	1.35	0.28	0.28	0.21	0.44	0.10
	30 DAT	1.35	0.83	0.84	0.47	0.66	0.41
Malshira	10 DAT	1.49	0.31	0.38	0.34	0.94	0.20
	30 DAT	1.19	0.66	0.49	0.22	0.31	0.35
Joldepa	10 DAT	1.23	0.43	0.56	0.34	0.96	0.18
	30 DAT	1.13	0.65	0.60	0.35	0.50	0.52
Sadapajam	10 DAT	1.24	0.51	0.39	0.35	0.74	0.11
	30 DAT	1.15	0.60	0.64	0.21	0.43	0.36
Lalpayka	10 DAT	1.18	0.39	0.32	0.17	0.95	0.13
	30 DAT	1.09	0.67	0.56	0.33	0.40	0.38

...Table 4.4 contd.

Table 4.4 contd. ...

Cultivars	Submergence at	Plant height	Tiller production	Shoot dry matter	Root dry matter	Crop growth rate	Relative growth rate
Mota	10 DAT	1.04	0.33	0.47	0.38	0.94	0.14
	30 DAT	1.07	0.59	0.47	0.20	0.24	0.13
Dudmona	10 DAT	1.26	0.47	0.46	0.62	0.94	0.23
	30 DAT	1.05	0.66	0.46	0.32	0.34	0.39
Jalkucha	10 DAT	1.68	0.51	0.35	0.35	0.84	0.24
	30 DAT	10.8	0.73	0.62	0.34	0.47	0.33
Lalchikon	10 DAT	1.24	0.32	0.46	0.50	0.95	0.23
	30 DAT	1.07	0.64	0.74	0.34	0.58	0.31
Achin	10 DAT	1.24	0.43	0.52	0.40	0.95	0.20
	30 DAT	1.02	0.65	0.84	0.46	0.71	0.52
Nakhuchimota	10 DAT	1.10	0.63	0.39	0.31	0.68	0.10
	30 DAT	1.18	0.67	0.77	0.45	0.63	0.41
Rajashil	10 DAT	1.23	0.39	0.35	0.35	0.65	0.13
	30 DAT	1.09	0.61	0.77	0.65	0.56	0.41
Kutiagni	10 DAT	1.35	0.51	0.52	0.41	0.95	0.23
	30 DAT	1.10	0.68	0.67	0.44	0.47	0.30
Sadamota	10 DAT	1.21	0.43	0.53	0.48	0.95	0.24
	30 DAT	1.04	0.78	0.79	0.72	0.57	0.42
Vog	10 DAT	1.29	0.61	0.56	0.40	0.98	0.29
	30 DAT	1.13	0.70	0.98	0.66	0.82	0.68

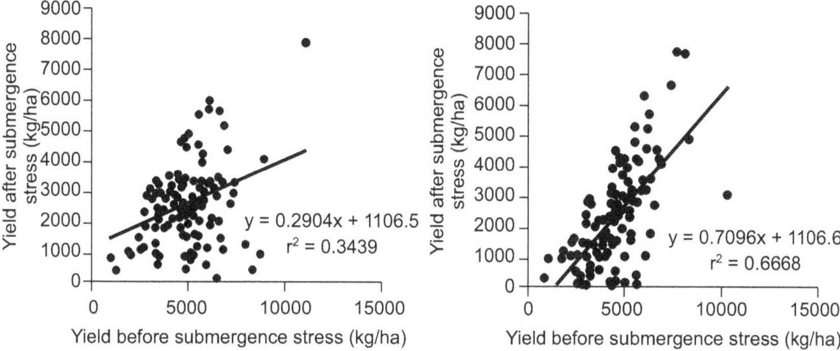

Fig. 4.14: Correlations of yields of rice among yield before submergence and the yield reduction due to submergence.

under waterlogged conditions, higher numbers were observed, ranging from 19 (Jeerigesanna) to 28 (BI33). Aerenchyma number showed highly significant positive correlation with pith area (r = 0.57) and negative correlation with xylem numbers (r = –0.46) and phloem number (r = –0.46). On the other hand, cross-section area radius positively correlated with pith area (r = 0.41) and xylem number (r = 0.70) (Fig. 4.14) (Kudur et al., 2015).

The major adaptive features of rice to waterlogging are the formation of aerenchyma. This aerenchyma constitutes the gas spaces and gets interconnected, which becomes the channel for continuous aeration between roots and shoots. In rice, aerenchyma is well developed in leaves, sheaths, roots, and internodes (Steffens et al., 2011; Pradhan et al., 2017). Generally, the formation of aerenchyma and its induction occurs within one to three days of anoxic treatment (Pradhan et al., 2017). In flood-water conditions, the presence of leaf gas films enhances the internal aeration and makes the rice tolerant to flood (Pedersen et al., 2009). The leaf gas film facilitates O_2 entry from surrounding water when in the dark, and CO_2 entry in light for photosynthesis. The role of gas films for flooding

Table 4.5: Average (±S.E.), minimum and maximum values of different traits of genotypes under normal and submerged conditions.

Trait	Control Av. (±S.E.) Min. Max.	Submergence Av. (±S.E.) Min Max.
1000 grain wt.(g)	23.24±1.47 17.31 37.89	18.79±1.26 2.98 36.28
Grains per panicle(g)	94±5.62 51 171	40±6.43 1 96
Potential yield (t/ha)	4.95±1.63 1.00 11.15	2.54±1.37 0.089 8
Plant height (cm)	105±5.32 84 132	121±4.92 100 151
Tillers per plant	15±0.955 10 24	- - -

Table 4.6: Estimates of per cent contribution of each character towards divergence in rice genotypes under submergence.

SI. No.	Characters	Contribution (%)
1	Absolute growth rate (mg/day/plant)	43.62
2	Days to 50 per cent flowering	22.87
3	Days to maturity	9.35
4	Plant height (cm)	8.76
5	Number of filled grains per panicle	4.17
6	Grain yield per plant (g)	4.17
7	Straw yield per plant (g)	2.30
8	Number of spikelets per panicle	2.21
9	1000 grain weight (g)	0.85
10	Number of tillers per plant	0.68
11	Harvest index (%)	0.60
12	Panicle length (cm)	0.34
13	Number of productive tillers per plant	0.09

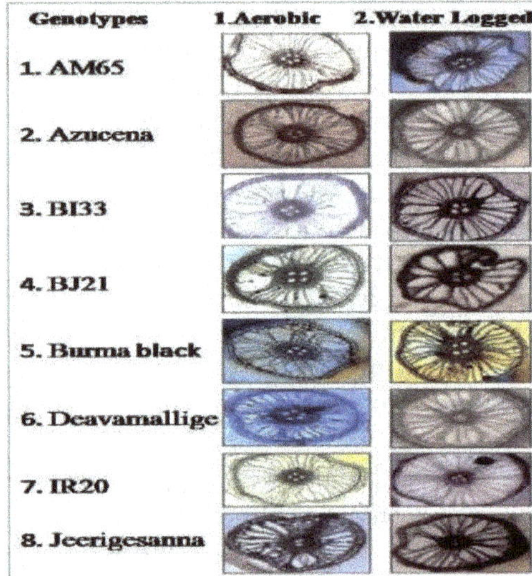

Fig. 4.15: Cross-sections depicting the anatomical differences between aerobic and waterlogged genotype.

tolerance has been confirmed through several studies by artificially-induced variation or loss of function mutants (Pedersen et al., 2009). Kurokawa et al. (2018) reported that 'Leaf Gas Film1' (LGF1/OsHDS1) enhances flooding tolerance in rice (Fig. 4.14).

The occurrence of submergence in agricultural land is increased by climate change. The level of tolerance in crops depends on many factors, including plant stages, and duration of submergence. Level of crop tolerance to submergence is likely to decrease when other adverse factors, such as salinity, exist. Two experiments have been conducted to study the effect of combined submergence and salinity on rice growth. First experiment studied the salt tolerance at early seedling stage while submerged. Split plot design was used with oxygen concentrations (with and without aeration) as the main plot and salinity levels (0.2, 50, 100 and 150 mM NaCl) as the sub plot. Second experiment evaluated the length of submergence (one and two weeks) and salinity levels (0.2 and 50 mM NaCl) at older stages of rice plants. The results showed that at early stages of seedlings, almost all growth variables decreased by salinity. The oxygen deficit reduced salt tolerance in rice seedlings. At older stages of rice, one-week submergence did not reduce rice growth. However, for two weeks' submergence, rice submerged in saline water showed better growth compared to those submerged in non-saline water (Tables 4.7 & 4.8 and Figs. 4.15 & 4.16) (Talpur, 2013).

In a quest for submergence tolerance at the seedling stage, Yamauchi et al. (1993) screened 258 accessions from the International Rice Research Institute (IRRI) gene bank and 404 from the International Network for Genetic Evaluation of Rice (INGER), using seeds pre-germinated for two days and then sown at 25-mm soil depth and submerged in 20–50 mm of water. Using this system, 12 genotypes were identified as tolerant, with emergence in the range of 54–78 %, compared with 7–19% for the sensitive genotypes. Furthermore, these authors observed that tolerant genotypes produce longer coleoptiles under hypoxia, in a manner independent of O_2 and ethylene concentrations; however, their mesocotyls and shoots elongate faster in response to ethylene than those of sensitive genotypes (Yamauchi et al., 1994; Yamauchi and Winn, 1996; Biswas and Yamauchi, 1997). Ling et al. (2004) evaluated 359 accessions from different sources, including indica and japonica accusing a water depth of 20 cm at 30°C. They used shoot elongation (coleoptiles) as the criterion for selecting tolerant accessions and identified reasonable variation in coleoptile length after five days under these conditions, although all genotypes failed to produce visible roots under such conditions. They further selected two contrasting genotypes, one with slow shoot elongation (DV85; 0.3 cm) and the other with faster shoot elongation (Kinmaze; 3.1 cm), and crossed them to develop a set of 81 recombinant inbred lines. These lines were subsequently genotyped to construct a genetic map, and phenotyped using the same criteria, leading to the detection of five quantitative trait loci (QTL), one each on chromosomes 1, 2 and 7, and two on chromosome 5. The method used for screening in these studies can detect variation in the ability of the coleoptiles to elongate under flooded conditions, but not

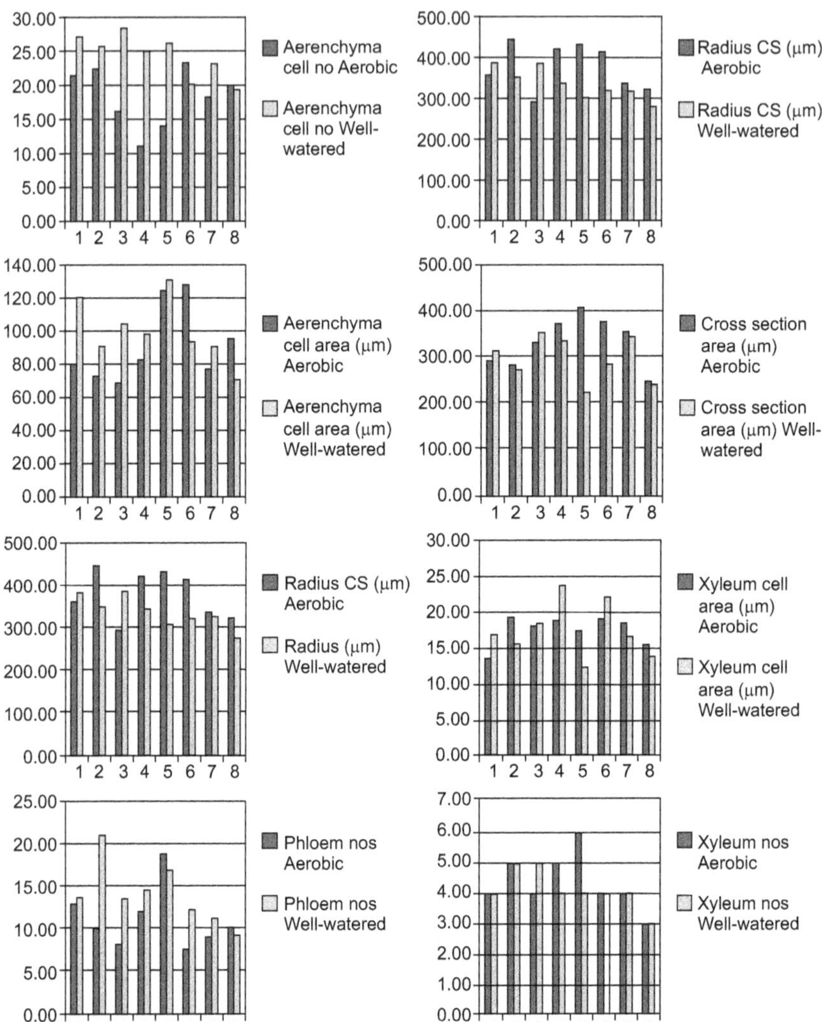

Fig. 4.16: Graphical representations of different root anatomical traits in two treatments (aerobic and well-watered) Note: Entries from 1 to 8 such as 1. AM65, 2. Azucena, 3. BI33, 4. BJ21, 5. Burmabalck, 6. Devamallige, 7. IR20 and 8. Jeerigesanna.

necessarily variation in ability to produce shoots and roots and survive the shallow water depths commonly experienced in flooded fields.

Recently, a small number of rice genotypes with greater tolerance of flooding during germination and early seedling growth were identified following large-scale screening of 8,000 gene-bank accessions and breeding lines at IRRI (Ismail et al., 2009; Angaji et al., 2010). These accessions were phenotyped by sowing dry seeds in soil and then flooding with

Table 4.7: Germination rate (%), vigor index, seedling height (cm), number of leaves and root length (cm) at 14 days after sowing.

Germination Rate (%)	Vigor Index	Seedling Ht. (cm)	No. of Leaves	Root Length (cm)
Aerated 99.68p	67.06p	19.50p	2.37p	8.60p
No aeration 99.76p	60.15p	11.34q	1.02q	4.37q
0 mMNaCl 99.84a	68.03a	18.33a	2.06a	10.35a
50 mMNaCl 100.00a	64.69ab	17.98a	1.78ab	6.90b
100mMNaCl 100.00a	60.31b	15.25b	1.61ab	4.47c
150mMNaCl 99.04b	61.37b	10.13c	1.33b	4.25c
Average 99.72	63.60	15.42	1.69	6.49
CV(%) 0.42	7.57	12.85	26.14	26.37
Interaction -	-	-	-	-

* Different letters indicate significant differences at P = 0.05 (DMRT)

** Symbol in parentheses indicates interaction between treatments

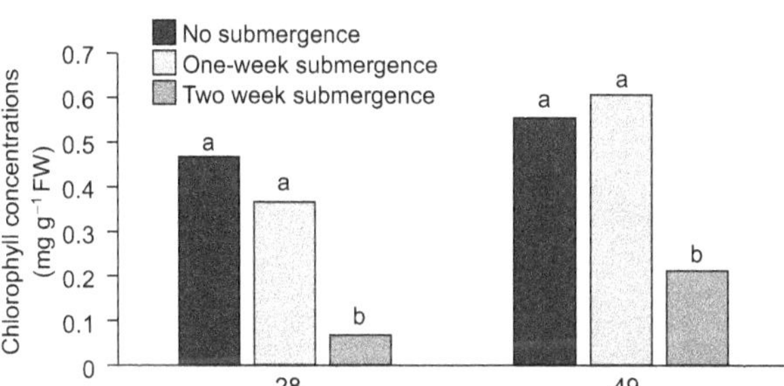

Fig. 4.17: Chlorophyll concentrations at a combined submergence and salinity stresses on 49 days after planting.

Table 4.8: Flowering age (days after planting) at a combined submergence and salinity stresses on 49 days after planting.

Submergence	Flowering Age (Days after Planting)	Salinity (dS m^{-1}) Average 0.3 4.9
No submergence	78.0c	85.7b 81.8
One week submergence	78.3c	85.3b 81.8
Two weeks submergence	0.0d	93.7a 46.8
Rerata	52.1	88.2 +
CV(%)	1.81	

* Different letters indicate significant differences at P = 0.05 (DMRT)

** Symbol in parentheses indicates interaction between treatment.

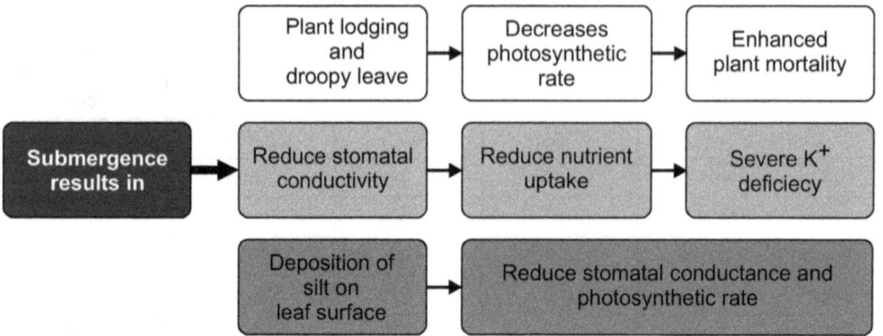

80–100 mm of water; this is more stringent than the screening methods of Yamauchi et al. (1993) using pre-germinated seeds and shallower water depth. Furthermore, tolerant genotypes were selected based on ability to generate roots and shoots, including new leaves, and to survive flooding by emerging from the floodwater within three weeks after sowing. In contrast, Ling et al. (2004) assessed only variation in coleoptile growth. One of these lines, 'Khao Hlan On', was crossed with a widely-grown lowland variety, 'IR64', and a BC2F2 population was developed and used for genetic analysis. Five QTLs were identified, one each on chromosomes

Table 4.9: Effect of submergence on shoot elongation (cm) and survival (%) in rice genotypes.

Parents	Shoot Elongation (cm)	Survival Percentage (%)
NDR-9830102	4.52	66.67
NDR-97	4.20	10.23
Swarna Sub1	5.23	85.64
Nagina-22	5.42	0.00
Mapping population		
1	3.12	70.00
2	4.00	60.00
3	3.86	74.00
4	2.80	85.00
5	3.50	75.00
6	4.20	65.00
7	5.40	20.00
8	2.56	90.00
9	3.45	72.00
10	4.00	68.00

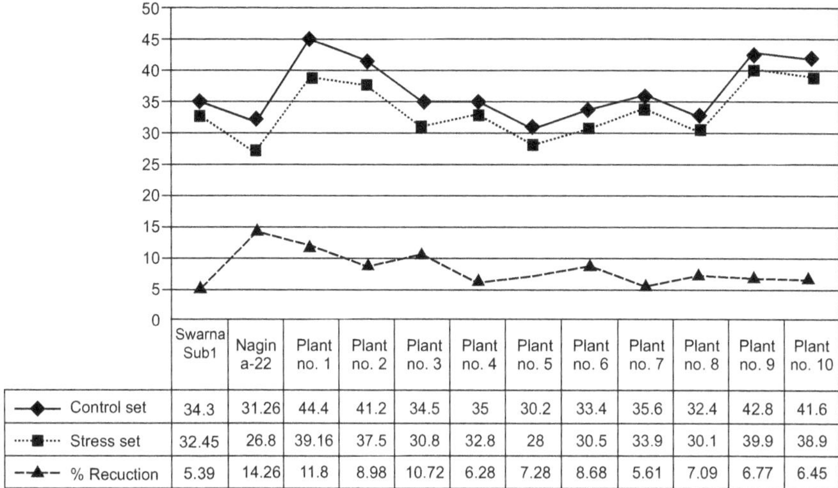

	Swarna Sub1	Nagin a-22	Plant no. 1	Plant no. 2	Plant no. 3	Plant no. 4	Plant no. 5	Plant no. 6	Plant no. 7	Plant no. 8	Plant no. 9	Plant no. 10
Control set	34.3	31.26	44.4	41.2	34.5	35	30.2	33.4	35.6	32.4	42.8	41.6
Stress set	32.45	26.8	39.16	37.5	30.8	32.8	28	30.5	33.9	30.1	39.9	38.9
% Recuction	5.39	14.26	11.8	8.98	10.72	6.28	7.28	8.68	5.61	7.09	6.77	6.45

Fig. 4.18: Chlorophyll content (SPAD value) variations under submergence and % reduction.

1, 3 and 7, and two on chromosome 9, explaining 18 to 34% of the phenotypic variation. Cultivar 'Khao Hlan On' contributed the tolerant alleles at all QTL loci. Two of these QTLs, on chromosomes 7 and 9, are considered major QTLs and are currently being fine-mapped and cloned to facilitate their use in breeding.

At the early stages, the germinating and growing embryo is dependent on stored carbohydrates in the seed endosperm and when absence of aerial or underwater photosynthesis prevents the generation of oxygen under water, and while aerenchyma tissue is still not well developed to provide aeration of the growing tissue. However, once seedlings become autotrophic, oxygen is made available either through underwater photosynthesis or direct contact with air. By then, the mechanisms associated with tolerance of flooding during germination will no longer dominate. Other mechanisms, such as rapid underwater elongation, as in deepwater rice varieties, or the dormancy strategy mediated by the Sub1 gene in submergence-tolerant varieties, become functional when the seedlings were flooded.

A pot experiment was conducted with Swarna Sub1 and Nagina 22 as parent and its mapping population BC2F3 (Swarna Sub1 X Nagina 22). The submergence treatment was given. Thirty-five-day-old seedling of rice was exposed to 14 days of complete submergence stress in the submergence tank of Department of Crop Physiology in Narendra Deva University of Agriculture and Technology, Kumarganj, Faizabad (U.P). Results are presented in Table 4.9 and Fig. 4.17

References

Abedin, M., Al-Mamun, M., Mia, M., and Karim, M. (2019). Evaluation of submergence tolerance in landrace rice cultivars by various growth and yield parameters. *J. Crop Sci. Biotech.*, 22: 335–344.

Alpi, A., and Beevers, H. (1983). Effects of O_2 concentration on rice seedlings. *Plant Physiology*, 71: 30–34.

Angaji, S.A., Septiningsih, E.M., Mackill, D.J., and Ismail, A.M. (2010). QTLs associated with tolerance of flooding during germination in rice (*Oryza sativa* L.). *Euphytica*, 172: 159–168.

Armstrong, W. (1979). Aeration in higher plants. *Advances in Botanical Research*, 7: 225–332.

Bailey-Serres, J., Fukao, T., Ronald, P.C., Ismail, A.M., Heuer, S., and Mackill, D.J. (2010). Submergence tolerant rice: SUB1's journey from landrace to modern cultivar, *Rice*, 3: 138–147.

Barik, J., Kumar, V., Lenka, S.K., and Panda, D. (2020). Assessment of variation in morpho-physiological traits and genetic diversity in relation to submergence tolerance of five indigenous lowland rice landraces. *Rice Sci.*, 27(1): 32–43.

Biswas, J.K., and Yamauchi, M. (1997). Mechanism of seedling establishment of direct-seeded rice (*Oryza sativa* L.) under lowland conditions. *Bot. Bull. Acad. Sin.*, 38: 29–32.

Bui, L.T., Ella, E.S., Dionisio-Sese, M.L., and Ismail, A.M. (2019). Morpho-physiological changes in roots of rice seedling upon submergence. *Rice Sci.*, 26(3): 167–177.

Colmer, T.D. (2003a). Aerenchyma and an inducible barrier to radial oxygen loss facilitate root aeration in upland, paddy and deep-water rice (*Oryza sativa* L.). *Annals of Botany*, 91: 301–309.

Colmer, T.D. (2003b). Long-distance transport of gases in plants: A perspective on internal aeration and radial oxygen loss from roots. *Plant, Cell & Environment*, 26: 17–36.

Colmer, T.D., and Pedersen, O. (2008). Underwater photosynthesis and respiration in leaves of submerged wetland plants: Gas films improve CO2 and O2 exchange. *New Phytologist.*, 177: 918–926.

Colmer, T.D., Armstrong, W., Greenway, H., Ismail, A.M., Kirk, G.J.D., and Atwell, B.J. (2014). Physiological mechanisms in flooding tolerance of rice: Transient complete submergence and prolonged standing water. *Progress in Botany*, 75: 255–307.

Colmer, T.D., and Voesenek, L.A.C.J. (2009). Flooding tolerance: Suites of plant traits in variable environments. *Functional Plant Biology*, 36: 665–681.

Crawford, R.M. (2003). Seasonal differences in plant responses to flooding and anoxia. *Canadian Journal of Botany*, 81: 1224–1246.

Dar, M.H., Chakravorty, R., Waza, S.A., Sharma, M., Zaidi, N.W., Singh, A.N. et al. (2017). Transforming rice cultivation in flood prone coastal Odisha to ensure food and economic security. *Food Secur.*, 9: 711–722.

Greenway, H., Armstrong, W., and Colmer, T.D. (2006). Conditions leading to high CO_2 (>5 kPa) in waterlogged-flooded soils and possible effects on root growth and metabolism. *Ann. Bot.*, 98: 9–32.

Han, C., Xiaojian,Y., He, D., and P. Yang. (2013). Analysis of proteome profile in germinating soybean seed and its comparison with rice showing the styles of reserves mobilization in different crops. *PLoS ONE*, 8: e56947.

Haque, M.Z. (1974). *Physiological behaviour of deep water rice.* Deep water Rice in Bangladesh. Bangladesh Rice research Institute. pp. 45–59.

Huo, W.H., Sun, F.Z., Peng, K.Q., Li, L.N., Wang, H.Q., and Xia, W. (1997). Effects of flooding on the yield and the yield components of rice. *Journal of Hunan Agricultural University*, 23: 50–53 (in Chinese with abstract in English).

Ismail, A.M., Ella, E.S., Vegara, G.V., and Mackill, D.J. (2008). Mechanisms associated with tolerance to flooding during germination and early seedling growth in rice (*Oryza sativa*). *Ann. Bot.*, 103: 197–209.

Ismail, A.M., Ella, E.S., Vergara, G.V. and Mackill, D.J. (2009) Mechanisms associated with tolerance to flooding during germination and early seedling growth in rice (*Oryza sativa*). *Ann. Bot.*, 103: 197–209.

Jackson, M.B., and Ram, P.C. (2003). Physiological and molecular basis of susceptibility and tolerance of rice plants to complete submergence. *Ann. Bot.*, 91(2): 227–241.

Jackson, M.B., and Armstrong, W. (1999). Formation of aerenchyma and the processes of plant ventilation in relation to soil flooding and submergence. *Plant Biology*, 1: 274–287.

Joshi, R., Vig, A.P., and Singh, J. (2013). Vermicompost as soil supplement to enhance growth, yield and quality of *Triticum aestivum* L.: A field study. *Int. J. Recycl. Org. Waste Agric.*, 2(1): 16.

Kato, Y., Collard, B.C., Septiningsih, E.M. and Ismail, A.M. (2014). Physiological analyses of traits associated with tolerance of long-term partial submergence in rice. *AoB Plants*, 6, pii: plu058.

Kawano, N., Ito, O., and Sakagami, J.I. (2009). Morphological and physiological responses of rice seedlings to complete submergence (flash flooding). *Annals of Botany*, 103: 161–169, pmid:18940854.

Kordan, H.A. (1974). The rice shoot in relation to oxygen supply and root growth in seedlings germinating under water. *New Phytol.*, 73: 695–697.

Kudur, P., Sanjay, R., Shasidhar, H., and Gowda, V. (2015).Study of rice (*Oryza sativa* L.) root anatomy under aerobic and waterlogged conditions. *Int. J. Applied & Pure Sci. and Agric.*, 1: 18-26.

Kurokawa, Y., Nagai, K., Huan, P.D., Shimazaki, K., Qu, H., Mori, Y. et al. (2018). Rice leaf hydrophobicity and gas films are conferred by a wax synthesis gene (LGF1) and contribute to flood tolerance. *New Phytologist*, 218: 1558–1569.

Lahari, G., B.M. Dushyanthakumar, G.B. Jagadeesh, G.K. Nishanth, and Raghavendra, P. (2017). Assessment of Genetic Diversity of Rice Genotypes for Submergence Tolerance in Rain-fed Lowlands. *Int. J. Curr. Microbiol. App. Sci.*, 6(11): 2149–2154.

Ling, J., Ming-Yu, H., Chun-Ming, W., and Jian-Min, W. (2004). Quantative trait loci and epistatic analysis of seed anoxia germinability in rice (*Oryza sativa*). *Rice Sci.*, 11: 238–244.

Magneschi, L., and Perata, P. (2009). Rice germination and seedling growth in the absence of oxygen. *Ann. Bot.*, 103(2): 181–196.

Matsukura, C., Saitoh, T., Hirose, T., Ohsugi, R., Perata, P., and Yamaguchi, J. (2000). Sugar uptake and transport in rice embryo. Expression of companion cell-specific sucrose transporter (OsSUT1) induced by sugar and light. *Plant Physiol.*, 124: 85–93.

Miro, B., Long kumar, T., Entilia, F.D., Kohli, A., and Ismail, A.M. (2017). Rice seed germination under water : Morpho-physiological responses and the bases of differential expression of alcoholic fermentation enzymes. *Front. Plant Sci.*, 8: https://doi.org/10.3389/fpls2017.01857.

Mohanty, S. (2013). Trends in Global Rice Consumption. *Rice Today*, 12: 44–45.

Mori, S., Fujimoto, H., Watanabe, S., Ishioka, G. and Kamei, M. (2012). Physiological performance of iron-coated primed rice seeds under submerged conditions and the simulation ofcoleoptiles elongation in primed rice seeds under anoxia, Soil Sci. *Plant Nutr.*, 58: 469–478.

Narsai, R., Wang, C., Chen, J., Wu, J., Shou, H., and Whelan, J. (2013). Antagonistic, overlapping and distinct responses to biotic stress in rice (*Oryza sativa*) and interactions with abiotic stress. *BMC Genomics*, 14: 93. Doi: 10.1186/ 1471-2164-14-93.

Ogiwara, H., and Terashima, K. (2001). A varietal difference in coleoptile growth is correlated with seedling establishment of direct seeded rice in submerged field under low-temperature conditions. *Plant Production Sci.*, 4: 166–172.

Pedersen, O., Rich, S.M., and Colmer, T.D. (2009). Surviving floods: Leaf gas films improve O2 and CO2 exchange, root aeration, and growth of completely submerged rice. *Plant J.*, 58(1): 147–156.

Pradhan, B., Kundu, S., Santra, A., Sarkar, M., and Kundagrami, S. (2017). Breeding for submergence tolerance in rice (*Oryza sativa* L.) and its management for flash flood in rain-fed low land area: A review. *Agric. Rev.*, 38(3): 167–179.

Puckridge, D.W., Kupkanchanul, T., Palaklang, W., and Kupkanchanakul, K. (2000). Production of rice and associated crops in deeply flooded areas of the Chao Phraya delta. *In: Proceedings of the International Conference: The Chao Phraya Delta: Historical Development, Dynamics and Challenges of Thailand's Rice Bowl*, 12–15 December, 2000, Kasetsart University, Bangkok, Thailand.

Ram, P.C., Singh, B.B., Singh, A.K., Ram, P., Singh, P.N., Singh, H.P. et al. (2002). Submergence tolerance in rain-fed lowland rice: Physiological basis and prospects for cultivar improvement through marker-aided breeding. *Field Crops Res.* 76: 131–15.

Raskin, I., and Kende, H. (1983). How does deep water rice solve its aeration problem? *Plant Physiol.*, 72: 447–454.

Raymond, P., Al-Ani, A., and Pradet, A. (1985). ATP production by respiration and fermentation, and energy charge during aerobiosis and anaerobiosis in twelve fatty and starchy germinating seeds. *Plant Physiology*, 79: 9879–9884.

Sarkar, R.K., Reddy, J.N., Sharma, S.G., and Ismail, A.M. (2006). Physiological basis of submergence tolerance in rice and implications for crop improvement. *Curr. Sci.*, 91: 899–906.

Sarkar, R.K., and Bhattacharjee, B. (2011). Rice Genotypes with Sub1 QTL Differ in Submergence Tolerance, Elongation Ability during Submergence and Re-submergence and Regeneration Growth at Re-emergence. *Rice*, 5: 7.

Sasidharan, R., and Voesenek, L.A. (2013). Lowland rice: High-end submergence tolerance. *New Phytologist*, 197: 1029–1031. Pmid:23373859.

Singh, A., Septiningsih, E.M., Balyan, H.S., Singh, N.K., and Rai, V. (2017). Genetics, physiological mechanisms and breeding of flood-tolerant rice (*Oryza sativa* L.). *Plant Cell Physiol.*, 58(2): 185–197.

Singh, S., Mackill, D.J., and Ismail, A.M. (2014). Physiological basis of tolerance to complete submergence in rice involves genetic factors in addition to the Sub1 gene. *AoB Plants*, 6: plu060.

Steffens, B., Geske, T., and Sauter, M. (2011). Aerenchyma formation in the rice stem and its promotion by H_2O_2. *New Phytol.*, 190(2): 369–378.

Talpur, M.A., Changying, J., Junejo, S.A., Tagar, A.A., and Ram, B.K. (2013). *African J. Agric. Res.*, 8: 4654–4659.

Toledo, A.M.U., Ignacio, J.C.I., Casal, C., Gonzaga, Z.J., Mendioro, M.S., and Septiningsih, E.M. (2015). Development of improved Ciherang-Sub1 having tolerance to anaerobic germination conditions. *Plant Breed. Biotechnol.*, 3: 77–87.

Tsuji, H. (1973). Growth and metabolism in plants under anaerobic conditions. *Environment Control in Biology*, 11: 79–84.

Turner, N. (1981). Techniques and experimental approaches for the measurement of plant water status. *Plant and Soil*, 58: 339–366.

Vergara, B.S., and Mazaredo, A. (1975). *Screening for Resistance to Submergence under Greenhouse Conditions;* Bangladesh International Rice Research Institute: Dhaka, Bangladesh, pp. 67–70.

Voesenek, L.A., and Bailey-Serres, J. (2013). Flooding tolerance: O_2 sensing and survival strategies. *Curr. Opin. Plant Biol.*, 16: 647–653.

Vriezen, W.H., Zhou, Z., and Straeten, D.V. (2003). Regulation of submergence-induced enhanced shoot elongation in *Oryza sativa* L. *Annals of Botany*, 91: 263–270. pmid:12509346.

Wang, B., Zhou, Y.J., Xu, Y.Z., Chen, G., Hu, Q.F., Wu, W.G. et al. (2014). Effects of waterlogging stress on growth and yield of middle season rice at the tillering stage. *China Rice*, 20: 68–72, 75.

Wati, I., Ehara, H., Kodama, I., Osawa, M., Suwignyo, R., Junaedi, A. et al. (2016). Growth characteristics of two Indonesian rice cultivars under several submergence durations, *Trop. Agric. Dev.*, 60: 40–47.

Winkel,A., Colmer, T.D., Ismail, A.M., and Pederson, O. (2013). Internal aeration of paddy field rice (*Oryza sativa*) during complete submergence-importance of light and flood water O2. *New Phytol.*, 197: 1193–1203.

Xiong, W., Asseng, S., Hoogenboom, G., Hernandez-Ochoa, I., Robertson, R.D. et al. (2019). *Nat. Food*, https://doi.org/10.1038/s43016-019-0004-2 (2019).

Xu, K., Xu, X., Fukao, T., Canlas, P., Maghirang-Rodriguez, R., Heuer, S. et al. (2006). Sub1A is an ethylene-response-factor-like gene that confers submergence tolerance to rice, *Nature*, 442(7103): 705–8.

Yamauchi, M., Aguilar, A.M., Vaughan, D.A., and Seshu, D.V. (1993). Rice (*Oryza sativa* L.) germplasm suitable for direct sowing under flooded soil surface, *Euphytica*, 67: 177–184.

Yamauchi, M., Herradura, P.S., and Aguilar, A.M. (1994). Genotype difference in rice post germination growth under hypoxia. *Plant Science*, 100: 105–113.

Yamauchi, M., and Winn, T. (1996). Rice seed vigor and seedling establishment in anaerobic soil. *Crop Science*, 36: 680–681.

5

Physiological Aspects

Generally, two types of flooding affect rice—one is, flash flooding (rapid rise of water levels with submergence of crop for one to two weeks) and another is stagnant flooding, where water level exceeds 100 cm depth and remains stagnant at these depths for several weeks. The semi-aquatic nature of rice plant makes it capable of growing under waterlogged and/or submerged condition for a considerable period. This is possible due to elongation of submerged shoot organ (Fig. 5.1).

The semi-aquatic nature of rice plant makes it capable of growing under waterlogged and/or submerged condition for a considerable period of time. This is possible due to elongation of submerged shoot growth at a faster rate by developing aerenchyma, that allows sufficient internal transport of oxygen to submerged plant parts from the re-emerged elongated shoot (Jackson and Ram, 2003; Magneschi and Perata, 2009). The process of energy generation and detoxification of fermented products through coleoptile elongation leads to survival under conditions of low oxygen (Miro et al., 2017). The germination and early seedling growth stage of rice has been found to be highly intolerant to submergence (Ismail et al., 2009; Angaji et al., 2010; Joshi et al., 2013). Apart from germination, submergence also brings many morphological and physiological changes in rice plant. During submergence, rice plant survives by elongation of leaf sheath and blade at seedling stage and internodes at vegetative growth stage. If the flooding duration exceeds two to three weeks, even the submergence-tolerant varieties try to expose their leaf tip above the water surface for ensuring survival (Sarkar and Bhattacharjee, 2011; Colmer et al., 2014).

Presence of hydrophobic wax on rice leaves creates a gas film, which helps with respiration and photosynthesis under submergence. Leaf Gas Film (LGF) reduced *in planta* accumulation of ethylene due to its better dissipation during submergence, thereby delaying the process

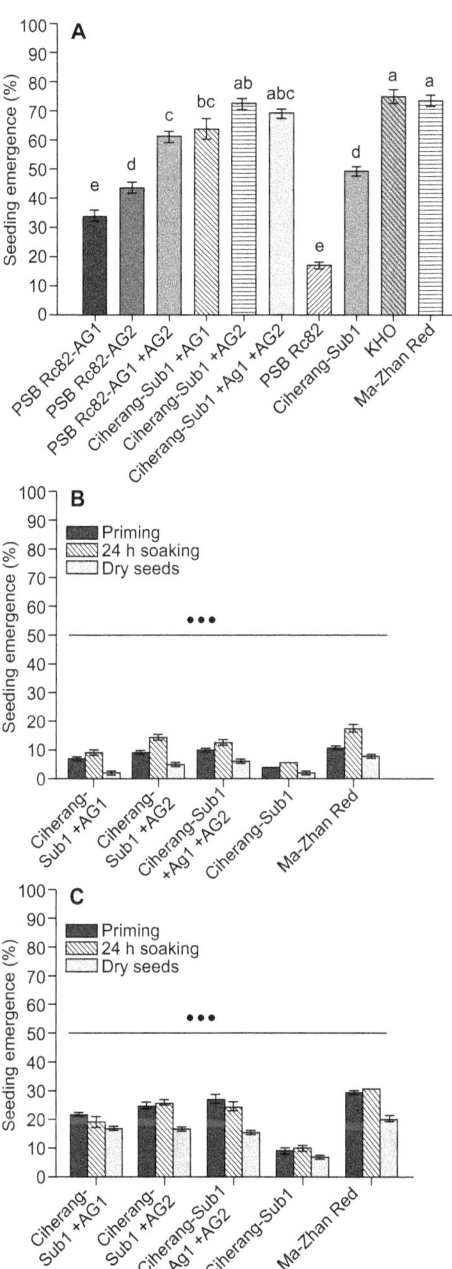

Fig. 5.1: Seedling emergence rate of AG lines under field condition with pre-treated seeds and flooding at 3–5 cm water depth, during the wet season of 2016 (**A–C**); (**A**) priming; (**B**) 24 h soaking; (**C**) dry seeds) and dry season 2017. Vertical bars indicate LSD 0.05 (n = 90).

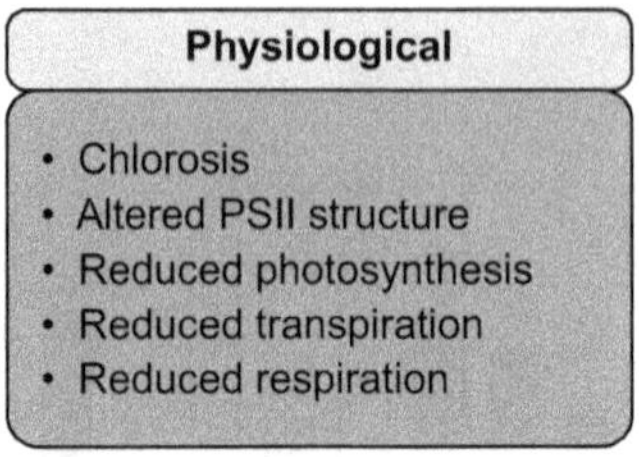

Physiological

- Chlorosis
- Altered PSII structure
- Reduced photosynthesis
- Reduced transpiration
- Reduced respiration

of ethylene-induced leaf senescence under submergence (Chakraborty et al., 2020). Complete submergence of plants prevents direct O_2 and CO_2 exchange with air. Underwater photosynthesis can result in marked diurnal changes in O_2 supply to submerged plants. Dynamics in pO_2 had not been measured directly for submerged rice (*Oryza sativa*), but in an earlier study, radial O_2 loss from roots showed an initial peak following shoot illumination.

In conclusion, underwater photosynthesis provides O_2 for internal aeration (Waters et al., 1989) as well as substrates for metabolism (Setter et al., 1987b) in submerged rice. The present study demonstrated marked dynamics in pO_2 within shoots and roots of submerged rice, in response to periods of darkness and light. Upon illumination, tissue pO_2 showed a rapid increase which then declined slightly to a new quasi-steady state, and this peak in pO_2 was transmitted from shoot to roots. The initial peak in pO_2 following illumination of submerged rice was likely to result from high initial rates of net photosynthesis, fuelled by CO_2 accumulated during the dark period.

Rice, in particular, has been shown to photosynthesize under water (Raskin and Kende, 1983; Setter et al., 1989) and rice grew well when submerged in water enriched with CO_2 to levels above air equilibrium to simulate some flood-waters (Pedersen et al., 2009; Setter et al., 1989). Like several other terrestrial wetland plants (Colmer and Pedersen, 2008b), rice possesses superhydrophobic, self-cleansing leaf surfaces that retain a thin gas film when immersed into water (Pedersen et al., 2009; Raskin and Kende, 1983; Setter et al., 1989). Leaf gas films markedly enhance gas exchange between leaf and flood-water so that underwater net photosynthesis (PN) is greater for leaves with gas films present, than when these are removed (Pedersen et al., 2009; Verboven et al., 2014; Winkel et al., 2013). In addition to carbohydrate production, underwater PN also results in better root aeration because much of the O_2 produced in the leaves diffuses via the aerenchyma down to the roots (Colmer and Pedersen, 2008a; Pedersen et al., 2009; Waters et al., 1989; Winkel et al., 2013). As O_2 production in underwater PN ceases at dusk, leaf gas films then also facilitate O_2 uptake from the flood-water, resulting in some internal aeration during darkness, but this is likely to be insufficient for the entire root system as root O_2

Fig. 5.2: Leaf gas film in submerged rice.

decreases to very low levels and fermentation occurs during dark periods (Pedersen et al., 2009; Waters et al., 1989; Winkel et al., 2013) (Fig. 5.2).

5.1 Net Photosynthesis

Measurements of underwater PN with 5 mol CO_2 m^{-3}, a level that saturates underwater PN of Swarna-Sub1 (irrespective of leaf gas films' presence or absence) in the present system (Winkel et al., 2013) was used to evaluate changes in capacity for underwater PN with time after submergence. All four genotypes had initial maximal underwater PN values ranging between 4.0 and 5.3 μmol O_2 m^{-2} s^{-1} (no significant difference). Capacity for underwater PN by FR13A and IR42 was significantly affected by time of submergence but maximal underwater PN of IR42 declined faster during the second week of submergence so that by the 13th day, the rate was only 9% of the initial capacity (Fig. 5.3a). Thus, during the latter part of the submergence treatment, capacity for underwater PN by FR13A was 6.7-fold higher than in IR42 (Fig. 5.3a). This superior performance of FR13A for retention of underwater photosynthetic capacity was not evident in Swarna-Sub1, which contains the SUB1 QTL from FR13A (Fig. 5.3b). The decline in capacity for underwater PN with time of submergence, in both Swarna-Sub1 and Swarna, was equal to that in IR42 (Fig. 5.3a, b). With high external CO_2 in the floodwater, PN under water was 13.4–19.5% of ambient rate in air. The lower PN rate under water than in air probably results from a combination of high resistance to gas exchange even in the presence of leaf gas films (Verboven et al., 2014) impeding O_2 exit that is further reduced by the relatively low solubility of O_2 in water, which would result in O_2 build-up inside the tissues, and thus high photorespiration under water, as previously discussed for rice by Setter et al. (1989).

Measurement of underwater PN with 0.2 mol CO_2 m^{-3}, a near ambient concentration in a similar field situation (Winkel et al., 2013) was used to evaluate field-relevant rates of underwater PN with time after

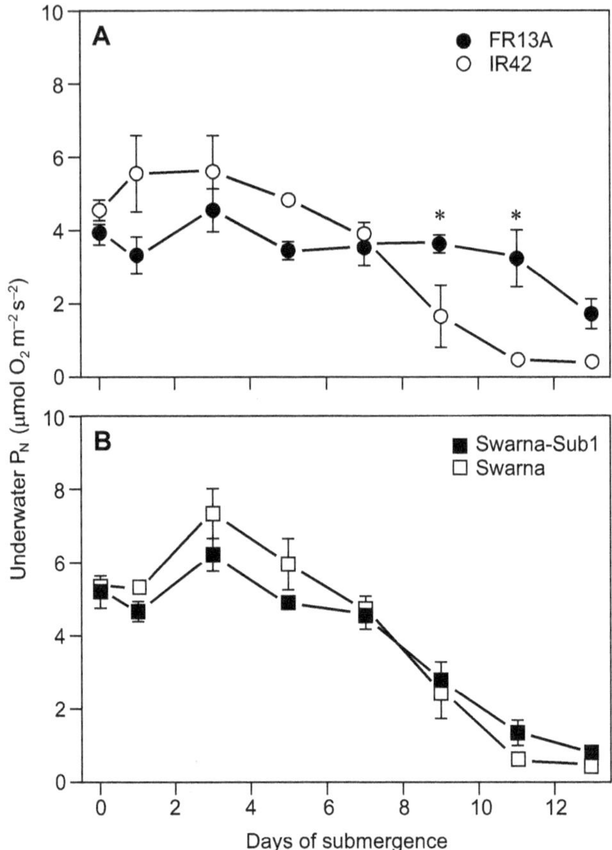

Fig. 5.3: Net photosynthesis under submergence.

submergence. At this CO_2 concentration, underwater PN is limited by CO_2 entry owing to the high resistance to diffusion from the bulk medium into the submerged leaf (Pedersen et al., 2009; Winkel et al., 2013). Therefore, gas film presence, a feature which reduces gas exchange resistance of submerged leaves (Colmer and Pedersen, 2008b; Raskin and Kende, 1983; Verboven et al., 2014), is of importance. Thus, the relationship of gas film persistence with underwater PN, and decline in leaf chlorophyll concentrations, both as influenced by time of submergence, are of importance to characterize contrasting genotypes.

All four genotypes initially possessed gas films on both leaf sides when submerged. These gas films were maintained near the initial thickness for the first four days in FR13A and IR42, and then declined with time of submergence (Fig. 5.4a, b). The decline, however, was initially faster for IR42 than FR13A, so that gas films were lost by the fifth day in IR42 and

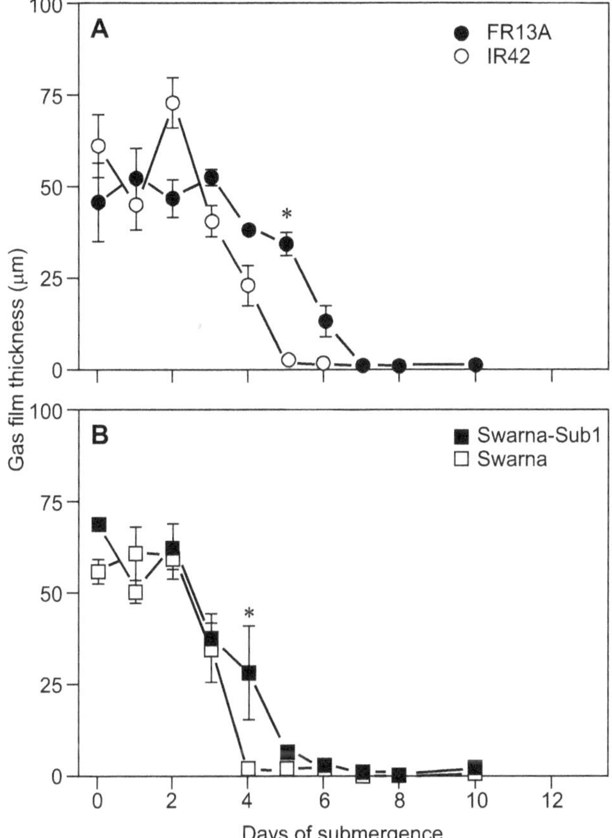

Fig. 5.4: Changes in gas film thickness (μm) in four rice varieties with increasing days of submergence.

by the seventh day in FR13A. The dynamics in the reductions in thickness of the gas films were, with exception of day 4, essentially the same for Swarna-Sub1 and Swarna.

Gas films on leaves of submerged wetland plants enable continued gas exchange via stomata and thus bypassing of cuticle resistance, enhancing exchange of O_2 and CO_2 with the surrounding water. Thus, during submergence, gas films on leaves enhance the photosynthesis and respiration of some wetland plants (Table 5.1).

5.2 Respiration

Thus, during submergence, gas films on leaves enhance photosynthesis and respiration of some wetland plants. Respiratory activities were compared

Table 5.1: Morphological, diffusion and respiration parameters of submerged leaves of rice (*Oryza sativa*); where values were not from own measurements, references are given in footnotes.

	Own Measurements (Mean ± Standard Deviation)	**Model**
Leaf morphology		
Lamina leaf thickness (μm)	116.0 ± 3.0	116.0
Porosity of lamina (%)	3.4 ± 1.6	5.0
Porosity of mesophyll tissue (%)		2.0
Epidermis thickness (μm)a		10.0
Stomatal density adaxial surface (mm^{-2})b	268 ± 64	239
Stomatal density abaxial surface (mm^{-2})b	294 ± 66	239
Axial distance between stomata (μm)	49.6 ± 6.2	49.0
Lateral distance between stomata (μm)	84.4 ± 13.6	84.0
Stomatal opening width, 100% open (μm)c		1.1
Stomatal opening width, 5% open (μm)c		1.1
Stomatal opening length, 100% open (μm)	16.3 ± 2.6	14.0
Stomatal surface area, 100% open (μm^2)		15.4
Stomatal surface area, 5% open (μm^2)		0.8
Area fraction stomata, 100% open (%)		0.37
Area fraction stomata, 5% open (%)		0.0185
Stomatal cavity radius (μm)a		13.0
Diffusion and respiration properties		
Diffusive boundary layer thickness (μm)	185	90/185/370
Gas film thickness (μm)	60	60
O_2 partial pressure (kPa)		2.5/5/10/21
O_2 concentration in air (mol m^{-3})		8.33/3.97/1.98/0.99
O_2 concentration in water (mol m^{-3})		0.288/0.137/0.069/0.034
O_2 diffusion coefficient in air at 30°C, $D^{4}_{O_{23}}$ ($m^2\ s^{-1}$)d		2.15×10^{-5}

among rice seedlings germinated in air for six days (aerobic seedlings), those germinated under water for five days (submerged seedlings), and those grown in air for a day after five days' submerged germination (air-adapted seedlings). The respiratory activity of the submerged seedlings increased rapidly on transfer to air and reached a plateau at 16 hours in air. Respiration of the submerged seedlings was as sensitive to cyanide as those of aerobic and air-adapted seedlings. 2,4-Dinitrophenol had no effect on the respiration of the submerged seedlings but stimulated those of the other two types of seedlings. Mitochondria from three types of seedlings did not differ in the ADP/O ratio and the respiratory control ratio (RCR) when succinate was oxidized. However, mitochondria from submerged seedlings (submerged mitochondria) showed poor RCR of unity when malate was oxidized. Both the rate of succinate oxidation and succinate dehydrogenase activity were low in submerged mitochondria, but increased during air adaptation. Although submerged mitochondria oxidized malate very slowly, the activity increased after exposure to air without any increase in malate dehydrogenase activity. When NAD+ was added to submerged mitochondria, oxidation of malate was restored to the level of the aerobic controls. Addition of NAD+ enhanced the state 3 rate in submerged mitochondria, and RCR recovered to nearly the same value as that of the aerobic controls. Similar effects of NAD+ on 2-oxoglutarate oxidation were observed. All these defects in submerged mitochondria were repaired during air adaptation. These results suggest that NAD+-linked substrate oxidation was low in submerged mitochondria because of NAD+ deficiency, and that oxidation increased with an increasing level of NAD+ during air adaptation (Fig. 5.5).

Various lines of 35 *O. sativa* and 27 *O. glaberrima,* including some classified as short-term submergence tolerant, were compared for submergence tolerance in field and pot experiments to long-term submergence-tolerant varieties in other words, deepwater varieties (Sakagami et al., 2009). Plants were submerged completely for 31 days in a field experiment, and partially or completely for 37 days in a pot experiment in a growth chamber. Leaf elongation and growth in shoot biomass during complete submergence in the field were significantly greater in *O. glaberrima* than in *O. sativa*. Submergence-tolerant cultivars of *O. sativa* were unable to survive prolonged complete submergence for 31–37 days, indicating that the mechanism of suppressed leaf elongation that confers increased survival of short-term submergence is inadequate for surviving long periods underwater. The *O. sativa* deepwater-adapted cultivar 'Nylon' and the 'Yele1A' cultivar of *O. glaberrima* succeeded in emerging above the flood-water. The photosynthetic rate was higher in deeply submerged plants than in non-submerged plants. The photosynthetic

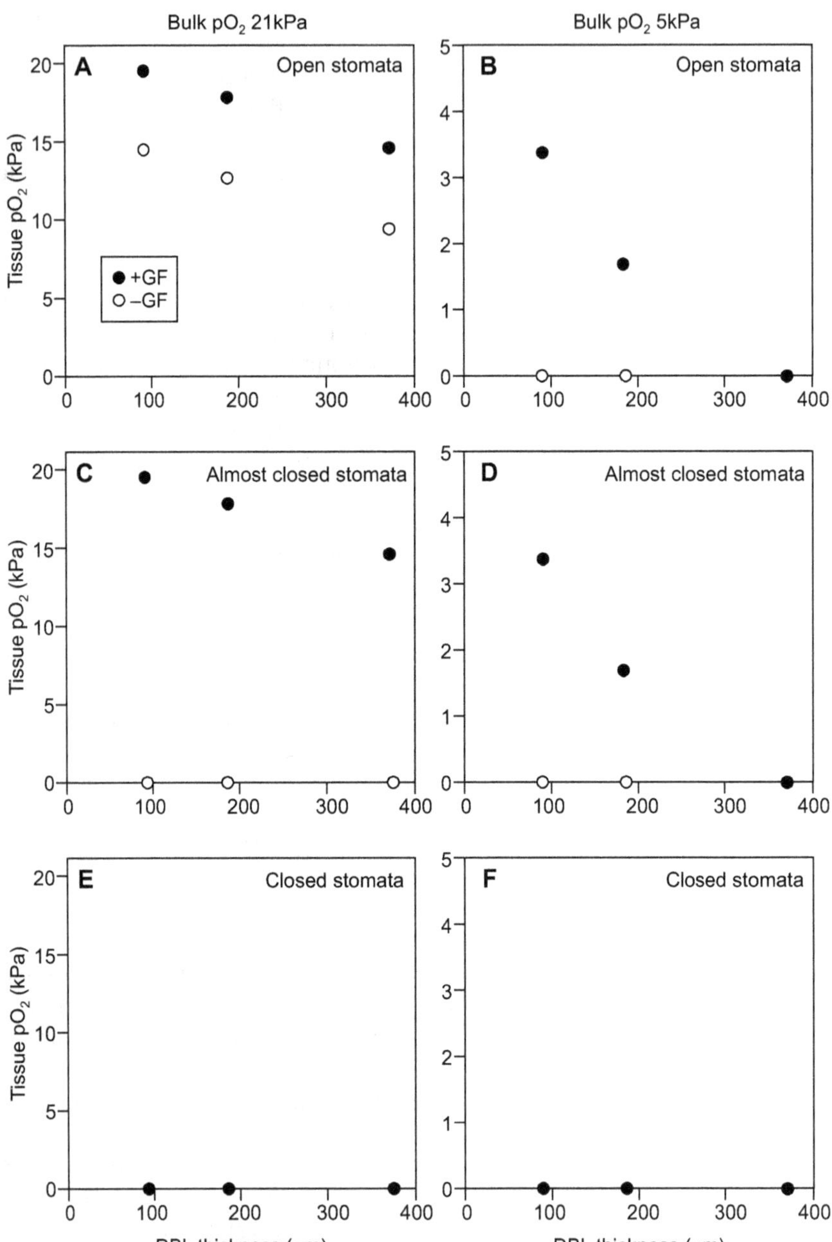

Fig. 5.5: Tissue pO_2 concentration in stomata with thickness of diffusion boundary layer.

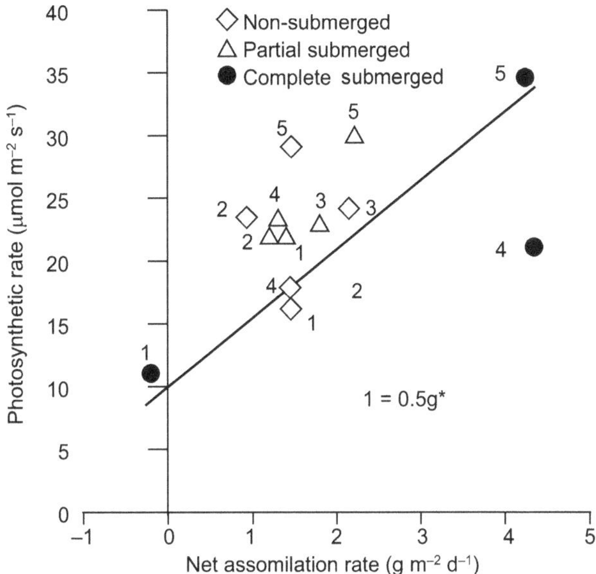

Fig. 5.6: Relationship between net assimilation rate during submergence and photosynthetic rate after 37-day submergence in a pot experiment. The number next each symbol indicate the cultivars: 1, Banjoulou; 2, IR71700; 3, IR73020; 4, Nylon; 5, Yele1A. Net assimilation rate indicates the increase of dry weight per unit area during 37-day submergence.

rate at 37 days after submergence in partial and complete submergence was closely related to the net assimilation rate during submergence (Fig. 5.6), which greatly increased shoot length, shoot biomass and leaf area, in association with an increased net assimilation rate compared with the lowland-adapted *O. sativa* 'Banjoulou'.

Photosynthesis in completely submerged wetland plants is severely impeded by low light, and the slow diffusion of CO_2 across the aqueous diffusive boundary layer (DBL) adjacent to leaves. Aquatic and amphibious plants have evolved a number of leaf traits to reduce the total resistance to CO_2 uptake, including: (i) dissected leaves, (ii) undulating leaf edges to increase turbulence across the leaf, (iii) thin leaves, (iv) reduced cuticle, and (v) chloroplasts in epidermal cells. The first three traits (i, ii and iii) serve to erode/reduce the DBL and thus decrease the distance of molecular diffusion (i.e., decreased total external resistance to CO_2 uptake), and (iv and v) reduce the resistance within the tissue for the diffusion pathway to chloroplasts. Likewise, many submerged terrestrial wetland plants also display some acclimation to inundation, such as thinner leaves, reduced cuticle, and chloroplast orientation towards the source of CO_2. Amphibious plants are positioned between the truly aquatic plants and the terrestrial wetland plants and display a suite of leaf acclimation traits that allow efficient gas exchange by leaves in air, as well as those formed under water.

Rates of net photosynthesis by submerged leaves are typically much lower than rates achieved in air, even for acclimated leaves. Underwater net photosynthesis by submerged terrestrial plants is generally lower than the rates achieved by aquatic plants. However, this is only true when photosynthetic rates are expressed on a per unit area basis, the units commonly used in terrestrial plant physiology. When net photosynthesis is expressed on a per unit dry mass basis, rates in aquatic plants > amphibious plants > terrestrial plants and this order reflects the higher carbon-return per unit of dry mass investment by the aquatic leaf types, as compared with terrestrial leaf types, when submerged.

Gas films on leaves of submerged terrestrial wetland plants have also been shown to facilitate underwater photosynthesis. Gas films form on hydrophobic leaf surfaces of many wetland plants when submerged, e.g., species of *Phragmites, Typha, Spartina, Carex, Phalaris* and *Oryza* (including cultivated rice), and the gas film forms a large gas-water interface that facilitates gas exchange with the surrounding water. It is likely that the stomata remains open underneath the gas film. The gas films enable leaves of such terrestrial wetland plants to photosynthesize under water, albeit at rates much reduced when compared with in air, but without further acclimation and this strategy may therefore be particularly advantageous under short floods; as examples, frequent tidal submergence or short duration flash floods, such as in some rice-growing areas and natural wetlands where water recedes after a week or two. The improved O_2 and sugar status of submerged rice owing to the beneficial effects of leaf gas films would enhance survival during complete submergence.

Due to enhanced synthesis and entrapment of ethylene under submerged conditions, the underwater leaf senescence is accelerated (Singh et al., 2014). Presence of hydrophobic wax on rice leaves creates a gas film, which helps with respiration and photosynthesis under submergence (Fig. 5.7 and Table 5.2).

5.3 Leaf Photosystem

Submergence-induced alteration of photosystem II (PS II) structure and function was probed using fast O-J-I-P chlorophyll a fluorescence transient and CO_2 photo-assimilation rate. Submergence resulted in an inhibition of CO_2 photo-assimilation rate and reduction in leaf chlorophyll content in rice, but the decrease was more in submergence susceptible (IR 42) and avoiding type (Sabita) cultivars compared to the tolerant (FR 13A) one. Quantification of the chlorophyll a fluorescence transients (JIP-test) revealed large cultivar differences in the response of PS II to submergence.

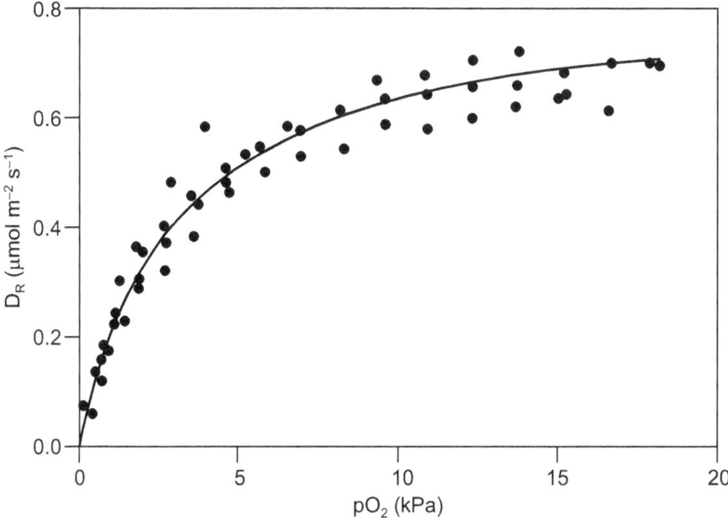

Fig. 5.7: *O. sativa* leaf lamina segments (gas films present) at various bulk water pO_2 at 30°C. The experimental data were fitted to a Michaelis-Menten model (V_{max} 0.83 ± 0.03 μmol m^{-2} s^{-1}; K_m 3.08 ± 0.037 kPa; r^2 0.91; error = SE). The maximum respiration rate corresponds to 0.012 mol O_2 m^{-3} s^{-1} expressed on a tissue volume basis using leaf dimensions given in Table 5.3.

Table 5.2: Changes of minimum (Fo), maximum (Fm) fluorescence and potential photochemical activity of PS II (Fv/Fm) due to submergence; the unit of each parameter is arbitrary.

DAS = days after submergence.

DAS	F_0			F_m			Fv/Fm		
	FR 13A	IR42	Sabita	FR 13A	IR42	Sabita	FR 13A	IR42	Sabita
0	508	520	500	2440	2520	2310	0.791	0.792	0.788
2	493	510	502	2608	2732	2669	0.806	0.813	0.810
4	530	554	597	2799	2592	2544	0.810	0.786	0.763
6	514	489	594	2606	1978	1833	0.802	0.752	0.661
8	494	537	535	2644	1776	1632	0.813	0.695	0.691
10	501	451	457	2245	1292	1038	0.775	0.620	0.524

The kinetics of chlorophyll, a fluorescence transient showed complex changes in the magnitude and rise of O-J, J-I and I-P phases of fluorescence rise. Due to submergence, both donor and acceptor sides of PS II were damaged and electron transport perturbed, which resulted in the fall of CO_2 photo-assimilation rate. Submergence also affected overall grouping probability (PG) or energetic connectivity between PS II antennae. The fall of PG was greater in submergence susceptible and avoiding types of cultivar compared to the tolerant one. These indicate that chlorophyll

Table 5.3: Changes of leaf chlorophyll content (mg g^{-1}fw) and CO_2 photo-assimilation rate (μmol CO_2 $m^{-2}s^{-1}$) in rice under different days of submergence. DAS = days after submergence.

DAS	Chlorophyll FR 13A IR42 Sabita	CO_2 Photo-assimilation Rate FR 13A IR42 Sabita
0	1.60 1.40 1.37	21.7 19.3 18.8
2	1.61 1.39 1.37	9.0 4.5 4.3
4	1.48 1.11 1.18	5.3 4.3 3.0
6	1.34 0.99 1.02	6.2 3.5 3.4
8	1.14 0.58 0.86	3.5 0.7 0.6
10	1.07 0.36 0.36	3.1 0.6 0.4

fluorescence can be used to know the sensitivity of rice to submergence stress (Table 5.3).

Rice plants cope with flash floods using either an 'escape strategy' involving rapid shoot elongation or a 'quiescence strategy' involving survival underwater with minimal activity. To clarify the differences in the response of leaf photosynthesis properties to conditions during and after submergence, two rice cultivars were compared: a non-shoot-elongating cultivar IR 67520-B-14-1-3-2-2 (IR67520) and a shoot-elongating cultivar IR72442-6B-3-2-1-1 (IR72442). Twenty-three-day-old seedlings were submerged in 80-cm-deep water for 14 days. During submergence, the chlorophyll contents of the upper fullyexpanded leaf (fifth leaf) and newly developed leaf later (sixth leaf) and the maximal quantum yield of photosystem II (F_v/F_m) of the fifth leaf decreased earlier in IR72442 than in IR67520. In the submerged sixth leaf, F_v/F_m was higher in IR72442 than in IR67520 at early measurement. Although F_v/F_m of the sixth leaf in submerged IR67520 increased substantially from two days post-submergence, IR72442 decreased because of leaf chlorosis. Therefore, a non-shoot-elongating cultivar coped with submergence by inhibiting photo-damage and maintaining high chlorophyll content in the leaves. The shoot-elongating cultivar was able to maintain the photosynthetic capacity of the newly developed leaf during submergence by prompt reduction of chlorophyll and chlorophyll fluorescence in the leaf that developed before submergence.

Ethylene is accumulated in rice plants during submergence because gas diffusion is 10^4-fold slower in solution than in air (Armstrong, 1979). The accumulation of ethylene in the submerged plants (approximately 0.50 Pa above that of non-submerged plants) exceeded in the minimum partial pressure (approximately 0.001 Pa) is known to be effective in rice when supplied in a gas phase (Ku et al., 1970; Konings and Jackson, 1979). Additionally, CO_2 deficiency caused by complete submergence will

Biochemical

- Rapid carbohydrate consumption
- Shift from aerobic to anaerobic respiration
- Disruption of osmotic and ionic homeostasis
- Disturbed cellular ROS balance
- Enhanced ROS generation after desubmergence
- Oxidative damage
- Lipid peroxidation

promote the responses to ethylene because CO_2 is antagonist to ethylene (Burg and Burg, 1965).

The present study is to characterise the non-structural carbohydrate (NSC) status and its catabolism along with elongation growth in rice cultivars either possessing or not possessing the Sub 1 quantitative trait locus (QTL), i.e., Swarna and Swarna Sub1 exposed to seven days of complete submergence. During submergence, Swarna accelerated the rate of stem and leaf elongation and rapidly consumed NSC. In contrast, Swarna Sub1 consumed energy resources more slowly and maintained similar growth rate to that of non-submerged plants. Swarna Sub1 showed better utilisation of carbohydrate than that of Swarna by progressive induction of alcohol dehydrogenase, starch phosphorylase, and total α-amylase enzyme activity during submergence. Overall, submergence tolerance conferred by the Swarna Sub1 QTL is correlated with better maintenance and utilisation of NSC than that of Swarna.

5.4 Tissue Carbohydrate Fractions

Flash flooding adversely affects rice productivity in vast areas of rain-fed lowlands in South and Southeast Asia. Tolerant genotypes were identified and key traits, such as high levels of non-structural carbohydrates (NSC, starch and soluble sugars) and limited underwater elongation, were found to be associated with tolerance. In a study, scientists evaluated the role of NSC before and after submergence and shoot elongation in submergence tolerance using genotypes that contrast in initial NSC content and elongation ability during submergence. The traits were further manipulated using a growth promoter, GA_3, and a potent gibberellin synthesis inhibitor, paclobutrazol (PB). Submergence for 10 days resulted in higher mortality of IR42 (intolerant), followed by Sabita and Hatipanjari. The latter two cultivars have a high initial NSC similar to that of the tolerant cultivar, FR13A, but elongated to higher extent under water. Exogenous GA_3 enhanced underwater elongation and depletion of NSC and reduced survival. Conversely, PB suppressed elongation, improved the retention of NSC, and enhanced survival. FR13A is less responsive to PB but more responsive to GA, suggesting its inherently low GA concentration under submergence. Underwater elongation is associated with NSC consumed during submergence but not with NSC level before submergence. Seedling survival showed a stronger association with NSC maintained after submergence ($r^2 = 0.74$) than with NSC before submergence ($r^2 = 0.27$). This suggests that carbohydrates maintained after submergence, being the result of both initial level and the level used during submergence, is more important for survival.

Non-structural carbohydrate contents before and after submergence are important for providing energy needed for maintenance of metabolism during submergence and for non-regeneration and recovery of seedlings after submergence (Jackson and Ram, 2003; Sarkar,1998; Das and Sarkar, 2001). Generally, results demonstrated that elongation competes for essential energy supply, and reducing elongation can conserve carbohydrates and increase submergence tolerance. Non-structural carbohydrates before submergence were highest in the shoots of both the tolerant (FR13A) as well as in the non-tolerant elongating genotypes (Sabita and Hatipanjari). However, upon submergence, a remarkable depletion of soluble carbohydrates in the shoots occurred, only in Sabita and Hatipanjari, as a consequence of excessive elongation (Tables 5.5 and 5.6). Carbohydrate contents were exhausted even to a greater extent upon the application of GA_3 in the lines with initially higher stem NSC, but the response of stem elongation to externally applied GA was much higher in FR13A and IR42. Paclobutazol application dramatically enhanced the retention of carbohydrate contents in all cultivars, and with greater effect on the tolerant line, FR13A (Table 5.4).

Seedling survival after submergence is substantially dependent on the level of NSC remaining in shoots after submergence, where 74% of the variation in survival could be explained by variation in stem NSC. However, seedling survival also showed strong association with shoot elongation during submergence ($r^2 = 0.56$), and to a lesser extent with NSC before submergence ($r^2 = 0.27$; Table 5.5).

Non-structural carbohydrate (NSC) accumulation in submergence-tolerant rice cultivars (cv) was studied in six Indica rice [*Oryza sativa* (L.)] cv under control and simulated submerged conditions. Tolerant cultivars accumulated greater contents of NSC compared to the susceptible cultivars. Starch and total NSC content showed significant positive association with survival percentage. On the other hand, elongation due to submergence was significantly a negative association with survival. The CO_2 photosynthetic rate, chlorophyll content, maximum photochemical efficiency of PS II (Fv/Fm), and activities of Rubisco were not significantly different between tolerant and susceptible cv under control condition. The

Table 5.4: Shoot non-structural carbohydrates (%) before (BS) and after (AS) submergence and with or without the application of GA_3, paclobutrazol or a combination of GA_3 + paclobutrazol.

Treatment	FR 13A			Sabita			Hatipanjari			IR42		
	BS	AS	Ch	BS	AS	Ch	BS	AS	Ch	BS	AS	Ch
Control	15.2a	10.0b	5.2	14.7a	6.3b	8.4	15.6ab	6.6b	9.0	8.3a	4.2b	4.1
GA_3	13.8b	7.9b	7.3	13.5b	4.2d	10.9	14.2c	4.6d	11.0	7.7a	3.6c	4.7
PB	15.5a	12.2a	3.0	15.1a	10.8a	3.9	15.9a	10.9a	4.7	8.6a	6.1a	2.2
GA_3+PB	15.0a	8.8c	6.4	14.5a	5.1c	9.6	14.9bc	5.2c	10.4	8.2a	3.9bc	4.4

BS: before submergence, AS: after submergence, Change (%) = (BScontrol AS). In a column, means followed by a common letter are not significantly different at P = 0.05 using Duncan's multiple range test.

Table 5.5: Correlation coefficients for the associations among percentage shoot elongation, non-structural carbohydrates (NSC) before and after submergence, NSC used during submergence and % survival.

Character	NSC Before Submergence	NSC After Submergence	NSC Used during Submergence	% Survival
% shoot elongation	–0.01 ns	–0.64**	0.70**	0.76**
NSC before submergence		0.63**	0.49 ns	0.52*
NSC after submergence			–0.37 ns	0.86**
NSC used during submergence				–0.34 ns

*, **: Significance at $P < 0.05$ and $P < 0.01$, respectively; ns: non-significant

ADP glucose pyrophosphorylase (AGPPase) activity was significantly higher in the tolerant cv and was a positive association with starch/NSC, whereas fructose 1,6-diphosphatase (FDPase) activity was significantly higher in susceptible cv compared to tolerant cv and was a negative association with starch/NSC. Greater activities of AGPPase along with lower activities of FDPase might facilitate greater accumulation of NSC in tolerant rice cultivars (Tables 5.6 and 5.7).

The detrimental effects of submergence on physiological performances of some rice varieties with special reference to carbohydrate metabolisms and their allied enzymes during post-flowering stages have been documented. It was found that photosynthetic rate and concomitant translocation of sugars into panicles were both related to the yield. The detrimental effects of the complete submergence were recorded in generation of sucrose, starch, sucrose phosphate synthase, and phosphorylase activity in the developing panicles of the plants as compared to those under normal or control (i.e., non-submerged) condition. The accumulation of starch was

Table 5.6: NSC (sugar and starch) content and plant height in different rice cultivars (21-days-old) without submergence (C) and 8 days after submergence (S).

Variety	Sugar (mg gfw^{-1})		Starch (mg gfw^{-1})		Total NSC (mg gfw^{-1})		Plant Height (cm)	
	C	S	C	S	C	S	C	S
FR 13A	34.89b	30.89b	26.39a	22.19ab	61.19a	52.99b	31.59d	42.39b
Khoda	42.39a	34.49b	23.79a	20.19b	65.99a	54.49b	29.79d	39.99c
Khadara	42.19a	36.99ab	24.29a	19.99b	66.39a	56.89ab	32.29d	43.79b
Kalaputia	36.69ab	31.99b	25.99a	21.99ab	62.69a	53.89b	32.29d	40.39c
Sarala	36.79ab	24.99c	13.49c	9.79d	50.29b	34.59c	28.69e	51.39a
IR 42	31.89b	21.39c	14.29c	8.99d	45.69b	30.49c	26.99e	52.59a

Means in a column with a common letter are not significantly different at LSD*pB0.05

Table 5.7: Association of sugar, starch, and total NSC contents with different enzymes of carbohydrate metabolism, photosynthetic parameters, elongation (%), and survival (%) in different rice cultivars during vegetative stage.

Parameters	Sugar	Starch	Total NSC
Rubisco	–0.080 ns	–0.179 ns	0.164 ns
AGPPase	0.329*	0.859**	0.756**
FDPase	0.018 ns	–0.826*	–0.541*
Invertase	0.003 ns	–0.290 ns	–0.191 ns
SSyn	–0.171 ns	–0.309 ns	–0.342 ns
SPSSyn	–0.331 ns	–0.004 ns	–0.243 ns
Photosynthetic rate	–0.430*	–0.234 ns	–0.083 ns
Chlorophyll content	0.169 ns	–0.272 ns	–0.087 ns
Fv/Fm	–0.495*	–0.032 ns	0.255 ns
Elongation (%)	–0.523**	–0.945**	–0.992**
Survival (%)	0.372*	0.992**	0.870*

significantly lower in plants under submergence and that was correlated with ADP-glucose pyrophosphorylase activity. Photosynthetic rate was most affected under submergence in varying days of post-flowering and was also related to the down-regulation of Ribulose bisphosphate carboxylase activity. However, under normal or control condition, there recorded a steady maintenance of photosynthetic rate at the post-flowering stages and significantly higher values of Ribulose bisphosphate carboxylase activity. Still, photosynthetic rate of the plants under both control and submerged conditions had hardly any significant correlation with sugar accumulation and other enzymes of carbohydrate metabolism, like invertase with grain yield. Finally, plants under submergence suffered significant loss of yield by poor grain filling which was related to impeded carbohydrate metabolism in the tissues. It is evident that loss of yield under submergence is attributed both by lower sink size or sink capacity (number of panicles, in this case) as well as subdued carbohydrate metabolism in plants and its subsequent partitioning into the grains (Fig. 5.8).

Research on the submergence stress of rice has concentrated on the quiescence strategy to survive in long-term flooding conditions based on Submergence-1A (Sub1A). In the case of the ripening period, it is important that submergence stress can affect the quality as well as the survival of rice. Therefore, it is essential to understand the changes in the distribution of assimilation products in grain and ripening characteristics in submergence stress conditions.

It was confirmed that the distribution rate of assimilation products in grain decreased by submergence treatment. These results were caused

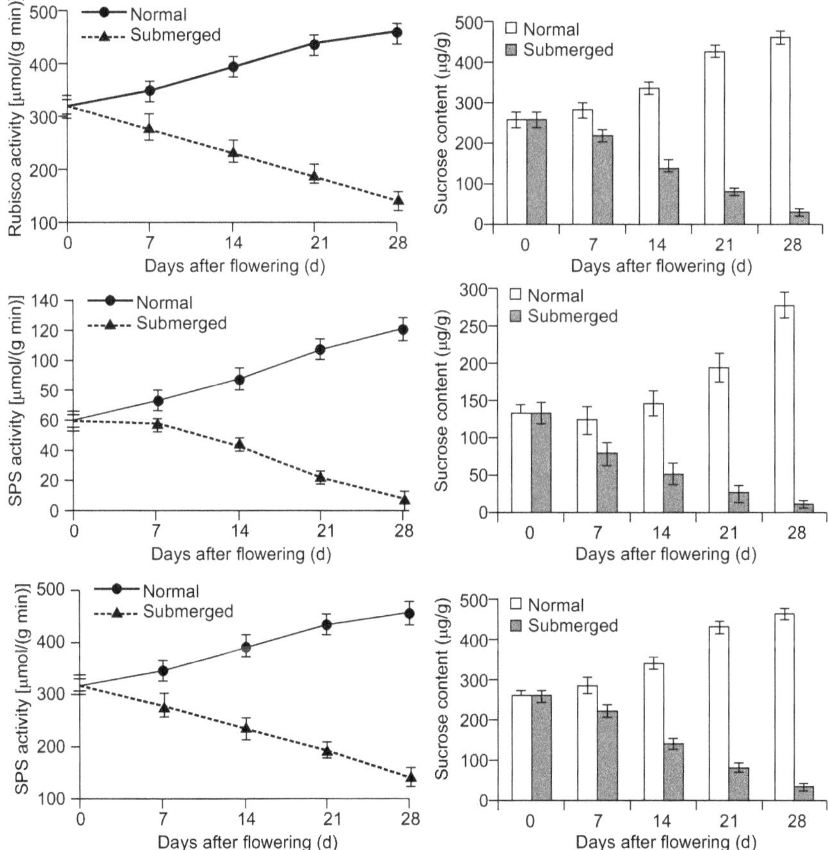

Fig. 5.8: Carbohydrate metabolic enzymes activities and starch content after flowering in normal and submerged rice.

by an increase in the distribution rate of assimilation products to the stem according to escape strategy. To understand this phenomenon at the molecular level, researchers analyzed the relative expression levels of genes related to sucrose metabolism, and found that the sucrose phosphate synthase gene (OsSPS), which induces the accumulation of sucrose in tissues, decreased in the seeds and leaves, but not in the stems. Furthermore, the sucrose transporter gene (OsSUT) related to sucrose transport decreased in the seeds and leaves, but increased in stems. Also analysed were the biological metabolic processes related to starch and sucrose synthesis, carbon fixation, and glycolysis using the KEGG mapper with selected differentially expressed genes (DEGs) in seeds, stems, and leaves caused by submergence treatment. It was found that the expression of genes for each step related to starch and D-glucose synthesis was downregulated in the seeds and leaves but up-regulated in the stem.

Although the distribution of assimilation products to grain increased, and decreased to stems and leaves, according to the ripening progress in the control, in the submergence treatment plots, the distribution of assimilation products to stems increased, resulting in an inhibited supply of assimilation products to grain (Fig. 5.10).

Recently, two groups of scientists clearly demonstrated that ethylene signalling pathways are a key factor for both submergence tolerance and escape strategies. In submergence-tolerant rice varieties, shoot elongation does not occur, and the plant consumes minimal carbohydrates because Sub1A, an ethylene response factor, represses GA responses (Xu et al., 2006; Fukao and Bailey-Serres, 2008). Contrarily, certain deepwater rice varieties undergo rapid inter-nodal elongation mainly by activation of SNORKEL1 (SK1) and SNORKEL2 (SK2), two other ethylene response factors that stimulate GA responses in intercalary meristem (Hattori et al., 2009).

Plant development and growth are mainly regulated by the integrated action of various hormones. In shoot apical meristem, for example, the integration of hormone signaling, such as that of cytokinins and GAs, is critical in maintaining meristematic identity (Kepinski, 2006). Likewise, at least three hormones (ethylene, ABA, and GAs) interact and integrate their signals to avoid submergence stress (Kende et al., 1998; Azuma et al., 2003; Voesenek et al., 2003; Xu et al., 2006).

FR13A, a submergence-tolerant cultivar, is able to survive complete submergence up to two weeks. It is known to acquire submergence tolerance by at least three separate processes. As the first process for submergence tolerance, the FR13A plant stops consuming carbohydrates upon submergence, resulting in energy preservation and stunted growth as mentioned above. With the saved energy, it resumes normal growth as flood-water is gone. De-submergence (exposure to air) after prolonged submergence rapidly increases reactive oxygen species (ROS), causing

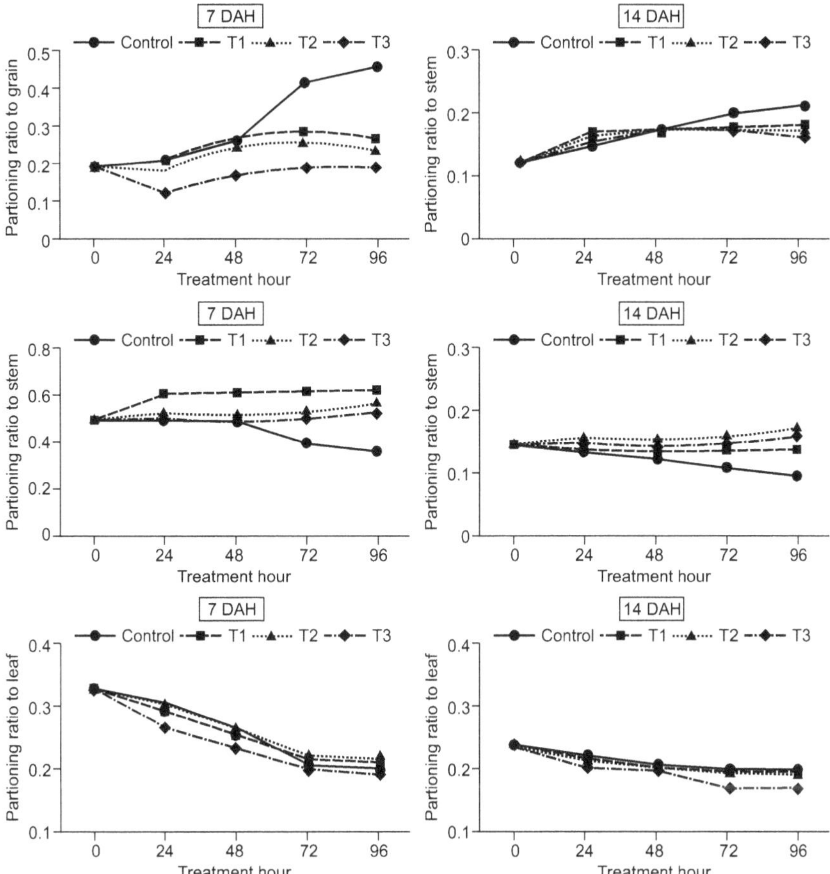

Fig. 5.9: Changes of distribution ratio of organs. The changes in the distribution ratio for organs (grain, stem, leaf) was affected by flooding treatment hour at seven, 14 days after heading (DAH). The treatment was carried out for up to 96 hours and the samples were used in five replicates. T1: over heading and clear water flooding condition. T2: flag leaf exposure and muddy water flooding condition. T3: over heading and muddy water flooding condition

fatal damage to submergence-sensitive species. To survive such oxidative stress during de-submergence, plants require antioxidant defence systems. Thus, it is plausible that high levels of antioxidant enzymes are active in submergence-tolerant lines during the process of de-submergence, while the enzyme levels are low in submergence-sensitive or submergence-avoiding lines or cultivars (Jung et al., 2010). As the second process, submergence-tolerant lines are well equipped with an antioxidant defence system which mitigates damage caused by elevated level of ROS during de-submergence process. Recently, it has been reported that ROS accumulation is not the only detrimental event occurring in

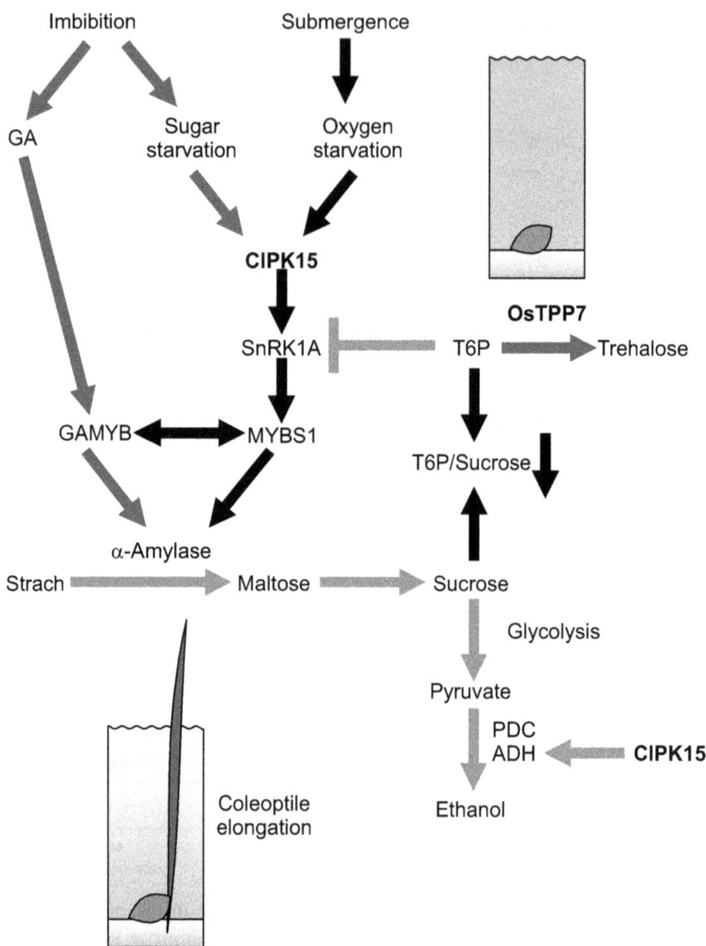

Fig. 5.10: Model of the mechanism of aerobic germination in rice. Red arrows indicate activation and blue lines indicate inhibition of aerobic germination. Gray arrows indicate common regulations in rice. Green arrows indicate the direction of the metabolic reaction.

de-submergence. Vegetative tissues, including leaves, have been found to undergo dehydration during de-submergence (Fukao et al., 2011). Finally, as the third process, submergence-tolerant lines also have increased responsiveness to ABA, leading to efficient acclimation to dehydration. In fact, ethylene is the first hormone to regulate initial physiological events, including intricate interaction with GA and ABA in submergence-tolerance response. In the case of submergence-escape response, a number of data regarding hormone metabolism and interaction have been accumulated. The main response to accomplish submergence escape is rapid elongation of stems or shoots. The physiology of submergence escape is well studied in deepwater rice. When deepwater rice is submerged partially, a continuous stream of air layers is formed throughout the leaf blades and acts as the major route for air supply (Raskin and Kende, 1983). This is possible because rice leaf blades are covered with hydrophobic epidermis which denies being soaked with water and eventually creates water-free zones between leaf surfaces and water. As the water level increases, gas composition inside internodal lacunae changes dramatically. Oxygen level rapidly decreases by up to 3% (v/v) to make lacunae hypoxic, while partial pressures of CO_2 and ethylene increase up to 6% (v/v) and 1 $\mu L\ L^{-1}$, respectively (Raskin and Kende, 1984; Stünzi and Kende, 1989). Maintenance of oxygen level in lacunae is attributable to photosynthetic activity and supply from air channels formed in leaf blades. Because of low diffusion coefficient of ethylene in water, the gas is trapped in lacunae while ethylene production continues, which leads to a rapid increase in ethylene concentration in intercalary meristem region. Increased levels of ethylene subsequently affect concentrations of two adverse hormones, ABA and GAs. In rice, ABA and GA act antagonistically as known in seed germination and coleoptiles' growth. This is the case in inter-nodal growth of deepwater rice (Fig. 5.11).

Growth of deepwater rice stems under submergence. Hypoxia occurs within 30 min. from the beginning of submergence. Increases in ethylene and GA levels subsequently follow hypoxia no later than two hours and four hours, respectively. Therefore, GA is considered to be the ultimate plant hormone that promotes elongation of deepwater rice internodes. GA is also known to induce ethylene production in deepwater rice internodes (Azuma et al., 1994), indicating synergistic interaction between ethylene and GA (Fig. 5.12).

When flash-flood-intolerant rice cultivars are submerged, they show greater morphological changes, such as elongation and chlorosis than tolerant cultivars. These morphological responses are caused by ethylene produced during submergence; however, a visible damage of intolerant cultivars is markedly developing after de-submergence rather than during submergence, which is probably due to oxidative damage. It was studied the effect of ethylene produced during submergence on antioxidant content and oxidative damage after de-submergence. When rice (*Oryza sativa*) was submerged for eight days, both tolerant cultivar (BKNFR) and intolerant cultivars (Mashuri and IR42) showed a decrease in ascorbate concentration during submergence. After three days of de-submergence, the tolerant cultivar showed a rapid recovery of total ascorbate and ascorbic acid, whereas intolerant cultivars showed a slow recovery of them, an increase in malondialdehyde formation, and low survival rate (about 30%). However, applying 200 mg l^{-1} of $AgNO_3$ as an ethylene antagonist to intolerant cultivars suppressed the decrease in ascorbate and increase in malondialdehyde formation after de-submergence, and

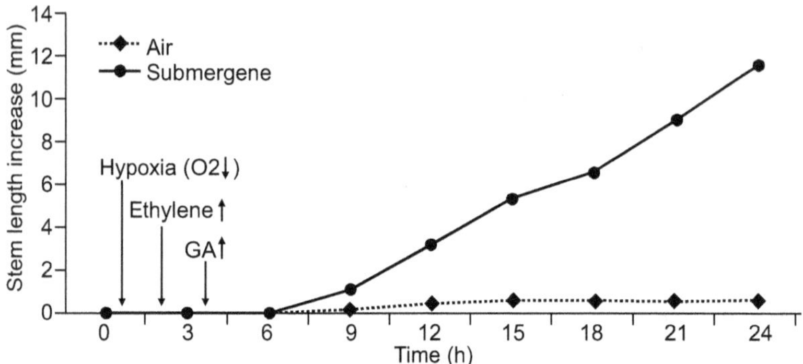

Fig. 5.11: Stem length increases under submergence with time.

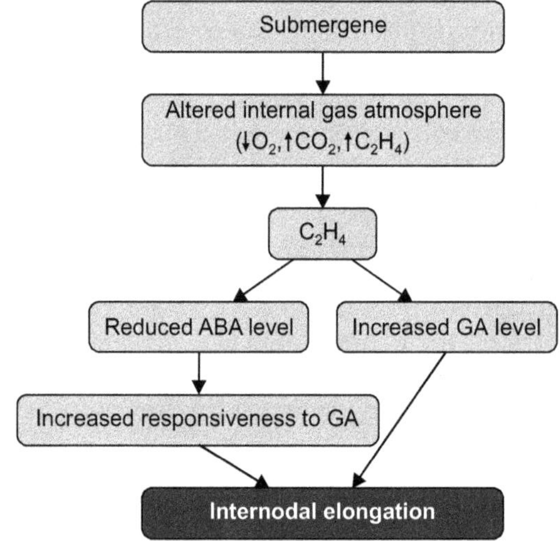

Fig. 5.12: Mechanism of internode elongation in submergence of rice.

improved survival rate to about 60%. Ascorbic acid supply to leaf discs from submerged IR42 suppressed increase in malondialdehyde formation by incubation under the light for 24 hours. In addition, strong negative correlations were observed between malondialdehyde formation with ascorbate concentration (r = −0.93) and with percentage of survival (r = −0.98). Results indicate that the accumulated ethylene during submergence adversely affected antioxidant mechanism in intolerant rice cultivars after de-submergence, and ascorbic acid was an important antioxidant *in vivo* for the recovery of submerged rice seedlings.

Deleterious effect of flood causes drastic reduction in rice production in low agricultural areas of Asia due to global climate change. Plants face oxygen deprivation under water and start low oxygen-dependent metabolic activity for proper utilization of carbohydrates with fermentative activities. In a study, leaf tissue of nine rice genotypes including *aus*-type submergence-tolerant variety FR13A and high-yielding submergence-sensitive line IR42 were used to establish a relationship between carbohydrate level and anaerobic enzyme activities. Three test sets for each accession were submerged for seven, 14 and 21 days for quantitative estimation of soluble and insoluble sugar content along with differential expression of three enzymes [pyruvate decarboxylase (*pdc1*), alcohol dehydrogenase (*adh1*) and sucrose synthase (*susy1*)] involved in carbohydrate metabolism in plants. From the findings it was observed that under submergence, FR13A showed higher insoluble sugar and lower soluble sugar along with higher expression of *pdc1* and *adh1*. Among the elongating categories, Kumrogarh showed the highest level of soluble and insoluble sugars with lower level of *pdc1,* and higher level of *adh1* and *susy1*. This promotes the Kumrogarh variety to lower the accumulation of cellular toxic acetaldehyde via ethanol production and increase the amount of sucrose. Keeping all in view, the Kumrogarh variety can be selected as submergence-tolerant deep-water line for lowland rice fields (Fig. 5.13).

Excess toxicity due to oxygen deficiency during natural submergence or flash flooding poses a potential hazard. Oxygen is essential for being alive for all the aerobic organisms. On the other hand, though the generation of reactive oxygen species (ROS) can affect both positive and negative roles, or that the ROS acts as an oxidative molecule, yet it has enough potential to cause toxicity in plants, such as rice (*Oryza sativa* L.) which is treated as semi-aquatic plant and is adapted to survive submergence for a considerable period of time. Further, rice can withstand submergence stress either by its inherent metabolic adaptations or by keeping its leaves above the water surface by continuously elongating the stem.

Using ethane as a marker for peroxidative damage to membranes by reactive oxygen species (ROS), it was examined the injury of rice seedlings during submergence in the dark. It is often expressed that membrane injury from ROS is a post-submergence phenomenon occurring when oxygen is re-introduced after submergence-induced anoxia. It was found that ethane production, from rice seedlings submerged for 24–72 hours, was stimulated to be 4–37 nl gFW^{-1}, indicating underwater membrane peroxidation. When examined a week later the seedlings were damaged or had died. On de-submergence in air, ethane production rates rose sharply, but fell back to less than 0.1 nl gFW^{-1} h^{-1} after two hours. Researchers compared submergence-susceptible and submergence-tolerant cultivars, submergence starting in the morning (more damage) and in the afternoon (less damage) and investigated different submergence durations. The seedlings showed extensive fatality whenever total ethane

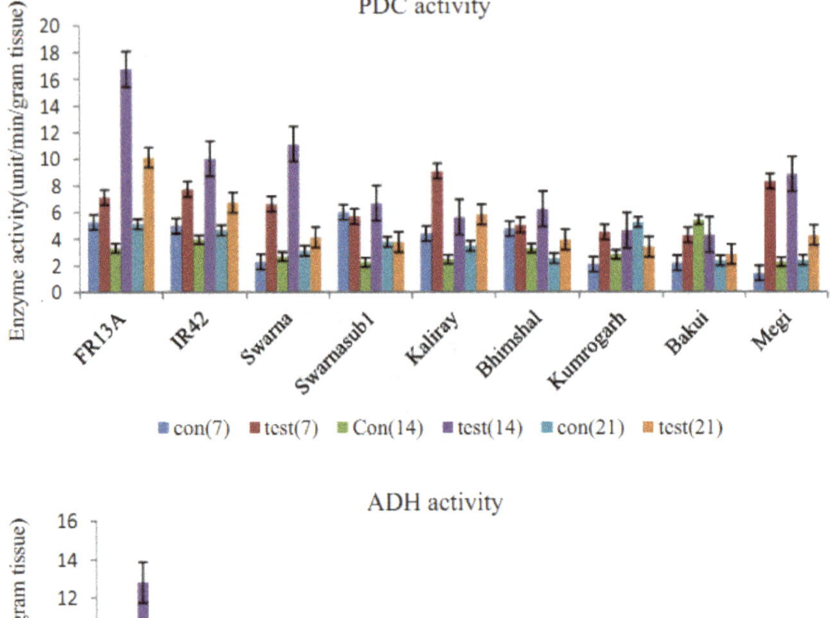

Fig. 5.13: PDC and ADH enzyme activities of rice varieties under different duration of submergence.

emission exceeded about 15 nl gFW^{-1}. Smaller amounts of ethane emission were linked to less extensive injury to the leaves (Figs. 5.14 and 5.15).

It is often expressed that membrane injury from ROS is a post-submergence phenomenon occurring when oxygen is re-introduced after submergence-induced anoxia. It was found that ethane production, from rice seedlings submerged for 24–72 hours, was stimulated to 4–37 nl gFW^{-1}, indicating underwater membrane peroxidation. The expression profile of pdc1, adh1 and susy1 genes in stressed and control plants are presented in Fig. 5.16a, b and c respectively. Flood-tolerant line FR13A showed

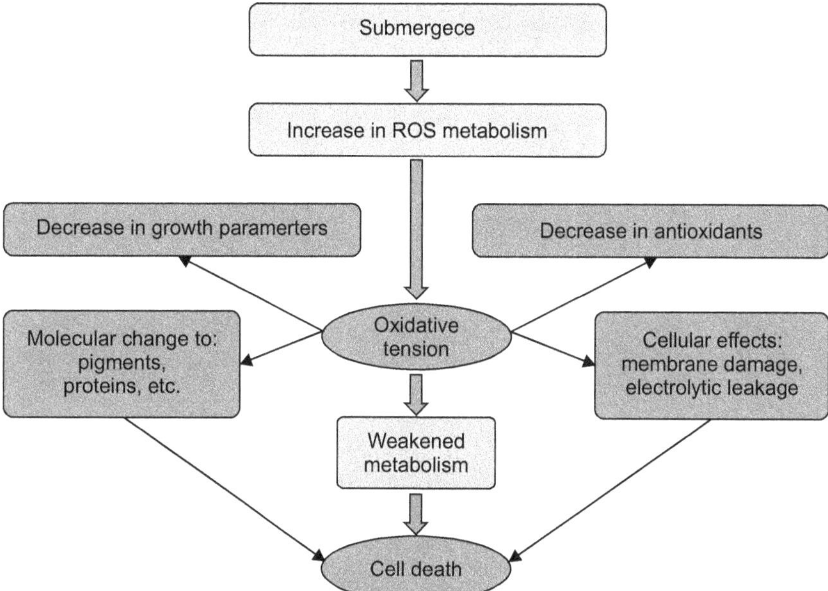

Fig. 5.14: Submergence increases the ROS production in cells. Prolonged conditions ultimately cause cell death (adopted from Upadhyay, 2018).

increased expression of pdc1 from 7th to 14th day interval under test conditions, which decreased almost twofold on 21st day of submergence. On the other hand, IR42, a sensitive line showed an increased level of pdc1 throughout the submergence period. Variety Kaliray showed higher expression of pdc1 in both control and submergence conditions but under submergence, the expression of pdc1 profoundly diminished along the three time points. No significant increase in the expression of pdc1 was found in var. Kumrogarh and var. Bakui under submergence, although on 14th day, pdc1 expression was significantly high in Kumrogarh. The expression of adh1 was found to be higher under submergence in comparison to control in all the cases. Interestingly, var. Kumrogarh showed the highest adh1 expression on 21st day of submergence, followed by Bakui. On the 14th day, tolerant genotype FR13A showed the highest expression of adh1. The sensitive IR42 showed comparatively lower level of expression of adh1 than the other line. Comparison of the transcript abundance of susy1 among the genotypes under submergence revealed that Kumrogarh, Kaliray and FR13A had higher expression than the other two genotypes. No apparent difference in gene expression was observed among the test and control sets for var. FR13A, IR42 and Kaliray. After 21 days of submergence, susy1 gene was found to be expressed at a lower level in both FR13A and Kaliray. Kumrogarh showed a progressively high

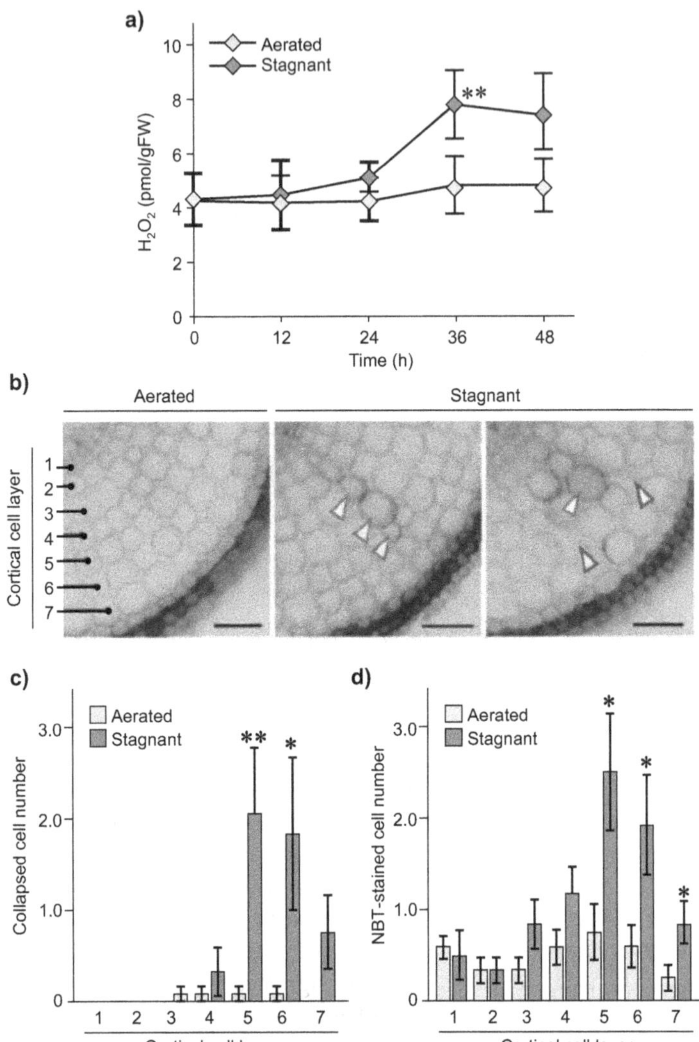

Fig. 5.15: Expression of cortical cells under normal and submerged rice.

expression of susy1 at the three time points. A higher expression of susy1 was also observed in var. Bakui on the 14th day, whereas, IR42 showed gradual reduction in the susy1 expression (Fig. 5.16).

- WRKY proteins are transcription factors (TFs) that regulate the expression of defence-related genes. The salicylic acid (SA)-inducible *Oryza sativa* WRKY6 (OsWRKY6) was identified as a positive regulator of *Oryza sativa pathogenesis-related 10a* (*OsPR10a*) by transient expression assays. A physical interaction between OsWRKY6 and

a

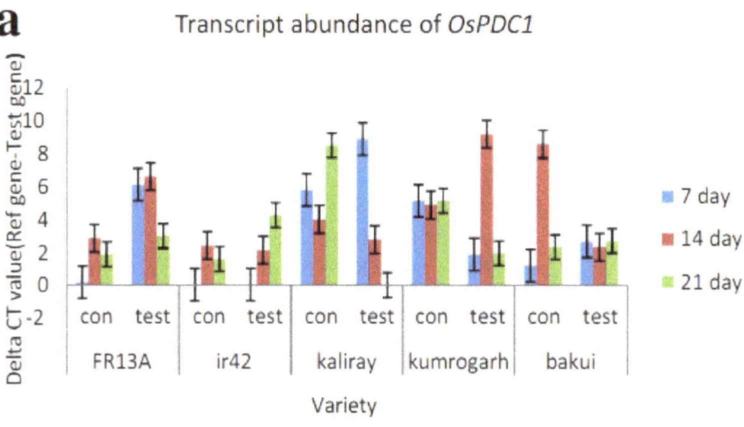

b

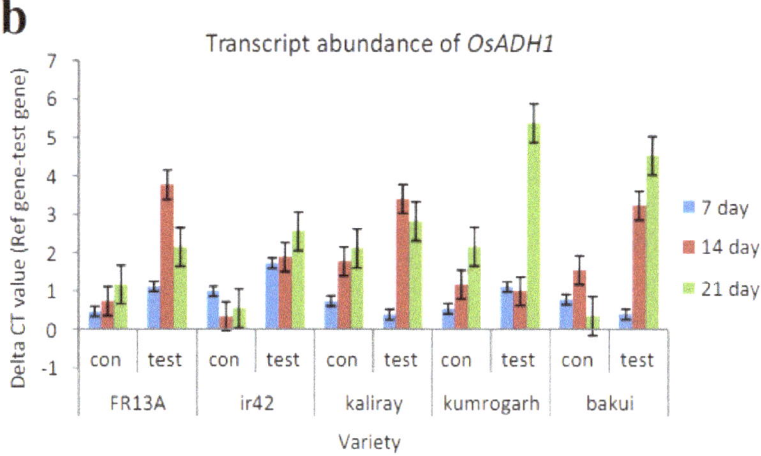

c

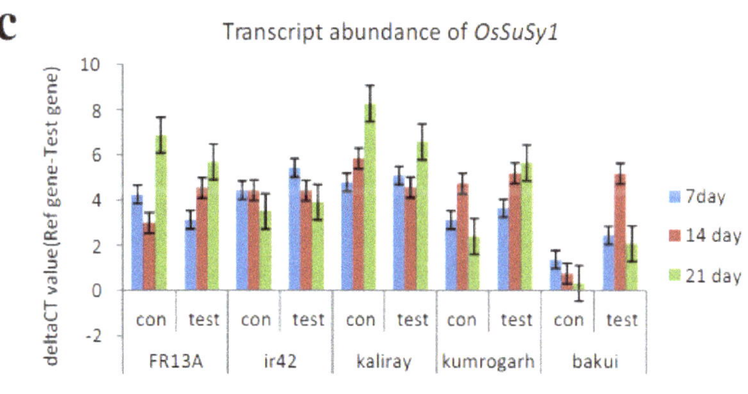

Fig. 5.16: Transcript abundance of rice carbohydrate metabolic enzymes.

W-box-like element 1 (WLE1), which positively regulates *OsPR10a/probenazole induced protein 1* expression, was verified *in vitro*. Several pathogenesis-related (*PR*) genes were constitutively activated, including *OsPR10a*, and transgenic rice (*Oryza sativa*) plants overexpressing (ox) *OsWRKY6* exhibited enhanced disease resistance to pathogens. By contrast, *PR* gene induction was compromised in transgenic *OsWRKY6*-RNAi lines, suggesting that OsWRKY6 is a positive regulator of defence responses. OsWRKY6-ox lines displayed leaf lesions, and increased OsWRKY6 levels caused cell death.

- Salicylic acid (SA) concentrations were higher in *OsWRKY6*-ox lines than in wild-type (WT) plants, and transcript levels of *Oryza sativa isochorismate synthase 1* (*OsICS1*), which encodes a major enzyme involved in SA biosynthesis, were higher in *OsWRKY6*-ox lines than in WT. OsWRKY6 directly bound to the *OsICS1* promoter *in vivo*. This indicates that OsWRKY6 can directly regulate *OsICS1* expression and thereby increase SA concentrations. OsWRKY6 autoregulates its own expression. OsWRKY6 protein degradation is possibly regulated by ubiquitination.
- Results suggest that OsWRKY6 positively regulates the defence responses through activation of *OsICS1* expression and OsWRKY6 stabilization (Fig. 5.17).

Submergence also stimulates the development of lysigenous aerenchyma, facilitating oxygen supply to the roots but the contribution of NO and RNS in aerenchyma formation under submergence is not known. Present study investigated the major components of the nitro-oxidative stress and their association with lysigenous aerenchyma development in the Sub1 near isogenic line of rice under submergence. Following submergence, Swarna showed increased NADPH oxidase (NOX) activity with excess reactive oxygen species (ROS) production in the roots. Submergence also caused increased NO content and membrane lipid peroxidation in Swarna roots. Submergence-induced ROS and RNS accumulation in roots disturbed the redox homeostasis leading to the formation of lysigenous aerenchyma through programmed cell death (PCD). PCD was also accompanied by altered cytoplasmic streaming and DNA damage. In the present study, Swarna Sub1 exhibited increased SOD, CAT, POX, APX, GR and GSNOR activity with subsequent detoxification of ROS and RNS; eventually decreasing the aerenchyma formation in the root under submerged conditions. Overall, the study established ROS and RNS-mediated unique mechanism in lysigenous aerenchyma formation in the rice roots under submergence.

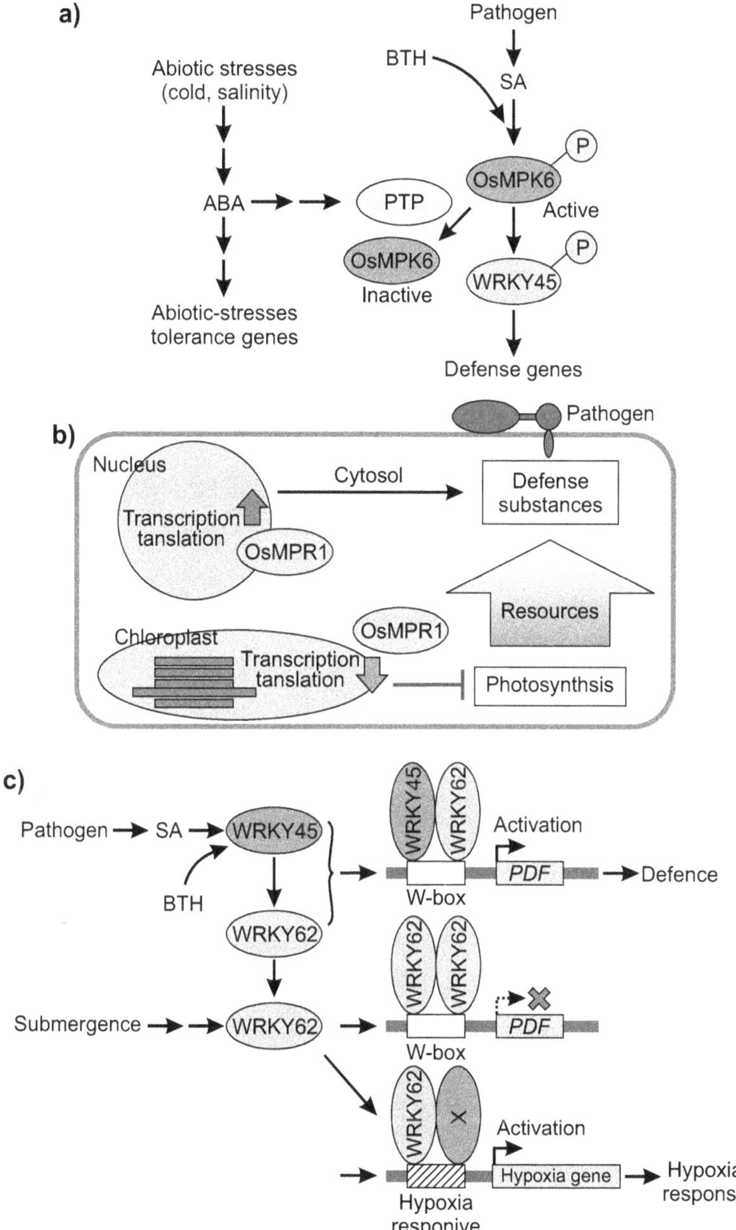

Fig. 5.17: Models for abiotic and biotic stress defence in rice.

Apoplastic ROS (superoxide; $^{\bullet}O_2^-$) production is mediated by the plasma membrane localized enzyme complex NADPH oxidase (NOX; EC 1.6.3.1) (Kaur et al., 2014). Consequently, the NOX respiratory burst oxidase homolog (RBOH) also contributes to the development of lysigenous aerenchyma under submergence (Yamauchi et al., 2017a). Elevated ROS eruption also triggers the programmed cell death (PCD) in the cortical cells, thereby regulating the lysigenous aerenchyma formation (Sarkar and Gladish, 2012). Plasma membrane degradation, nuclear disruption, and alteration of cytoplasmic streaming are the early indicators of PCD. This is later followed by the cytoplasmic streaming arrest and collapse of vacuolar membrane (Dauphinee et al., 2014).

Nitric oxide (NO) also plays a major role in regulation of lysigenous aerenchyma under hypoxic conditions (Wany et al., 2017). Studies on the physiological relevance of NO and other NO-related molecules, including S-nitrosoglutathione (GSNO) and peroxynitrite ($ONOO^-$), collectively termed as the reactive nitrogen species (RNS) under different abiotic stress in plants, has opened new avenues of research (Nabi et al., 2019). Several studies have revealed the nitrosative stress to change the cellular redox signalling, leading to PCD in plants (Wilkins et al., 2011). Interaction of NO and RNS with ROS has been extensively studied in plants under salinity (Jain et al., 2018), low temperature (Airaki et al., 2012), mechanical wounding (Houmani et al., 2018), and heavy metal stress (Terron-Camero et al., 2019). However, the contribution of RNS signalling in lysigenous aerenchyma development in rice roots under submergence is so far unexplored.

Redox homeostasis in rice under submergence is mediated by the antioxidant defence system (Huang et al., 2019). Several enzymatic and non-enzymatic antioxidants contribute in cellular detoxification of ROS/RNS in plants. First-line defence antioxidants include superoxide dismutase (SOD; EC1.15.1.1), catalase (CAT; EC 1.11.1.6), guaiacol peroxidase (POX; EC 1.11.1.7), and ascorbate peroxidase (APX; E.C.1.1.11.1). SOD catalyses the dismutation of $^{\bullet}O_2^-$ into hydrogen peroxide (H_2O_2) and oxygen (O_2) (Das and Roychoudhury, 2014), whereas, CAT, POX, and APX are responsible for eliminating H_2O_2 from different cell organelles. Glutathione Reductase (GR; E.C.1.6.4.2) reduces oxidized glutathione (GSSG) to GSH. The second line of defence comprising of non-enzymatic antioxidants, such as ascorbate and glutathione scavenge ROS by their reducing mechanism. The S-nitrosoglutathione reductase (GSNOR; EC 1.2.1.46) plays a crucial role in RNS homeostasis in plants by catalysing the reduction of GSNO to GSSG and ammonia. Several studies have shown the down-regulation of the ROS scavenging *metallothionein 2b* gene (*MT2b*) to promote the formation of lysigenous aerenchyma in the stem internodes of rice

(Steffens et al., 2013). Possible involvement of *Metallothionein1* genes in lysigenous aerenchyma formation in rice roots is also reported (Yamauchi et al., 2017). However, the regulatory role of the antioxidant enzymes during the aerenchyma formation has not been studied in rice roots.

ROS are identified as secondary class of smaller molecules which act as mediators to the responses of biotic and abiotic stresses, including flooding. ROS are capable of irreversible cellular damage by protein oxidation, enzyme inactivation, changes in gene expression, and decomposition of biological membranes, etc., and result in cell death.

Moreover, ROS generated during and following submergence are usually lower and the concentration of antioxidants (ascorbic acid, glutathione, phenolics and antioxidant enzymes) higher in tolerant landraces, contributing better sustainability.

Submergence-induced ROS accumulation was also determined in the rice roots by measuring the endogenous • O_2^- production rate and H_2 O_2 content. For instance, Swarna exhibited significantly increased • O_2^- production rate under submergence ($P \leq 0.05$), which increased with the duration of stress (63.9 and 72.2%) in submergence.

Rice cultivars effective in ROS detoxification following de-submergence can retain chlorophyll, support plant development, regenerate new leaves, and keep older leaves. As a result, **rice** plants protect themselves from oxidative damage via two mechanisms: (1) antioxidant enzyme systems and (2) natural antioxidants.

Plant cells are able to neutralise the ROS with efficient oxygen-scavenging machinery consisting of several antioxidants (both non-enzymatic and enzymatic) (Gill and Tuteja, 2010). Thus, induction of antioxidant defence mechanism is essential for protection of plants against different stresses. SOD, catalase (CAT), guaiacol peroxidase (GPX), ascorbate peroxidase (APX), glutathione reductase (GR), dehydroascorbate reductase (DHAR), and monodehydroascorbate reductase (MDAR) are the predominant antioxidant enzymes, which play important roles in protecting plants from oxidative damage.

The activities of CAT, APX, SOD, and POX in rice seedlings is under different levels of submergence. Enzyme activity was detected from detached shoots of 14 day-old rice seedlings after treated control (CK), partial submergence (PS), and full submergence (FS) for six days, then recovery one day (R1D). The catalase (CAT), ascorbate peroxidase (APX), superoxide dismutase (SOD), and total peroxidase (POX) activity were determined, respectively. The data represent average values ± SD from six biologically independent experiments. Values with different letters are significantly different at $P < 0.05$, according to post-hoc LSD (Figs. 5.18, 5.19 and 5.20).

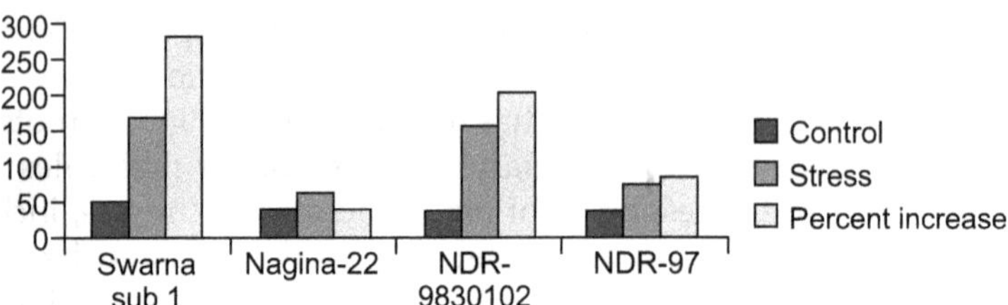

Fig. 5.18: Effect of submergence treatment on catalase activity (unit g^{-1} fresh wt.) of rice genotypes.

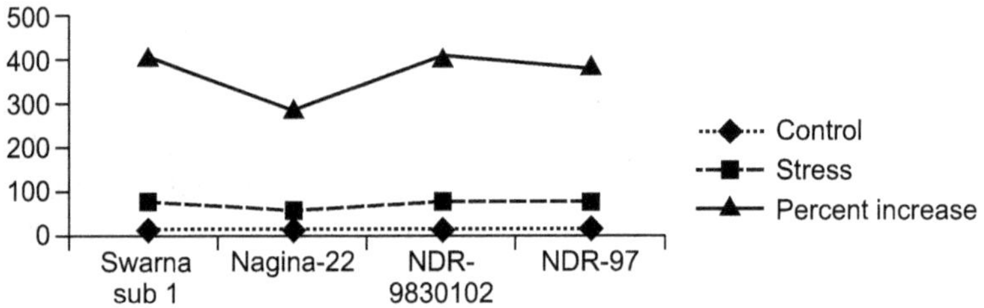

Fig. 5.19: Effect of submergence treatment on peroxidase activity (unit g^{-1} fresh wt. min^{-1}) of rice genotype.

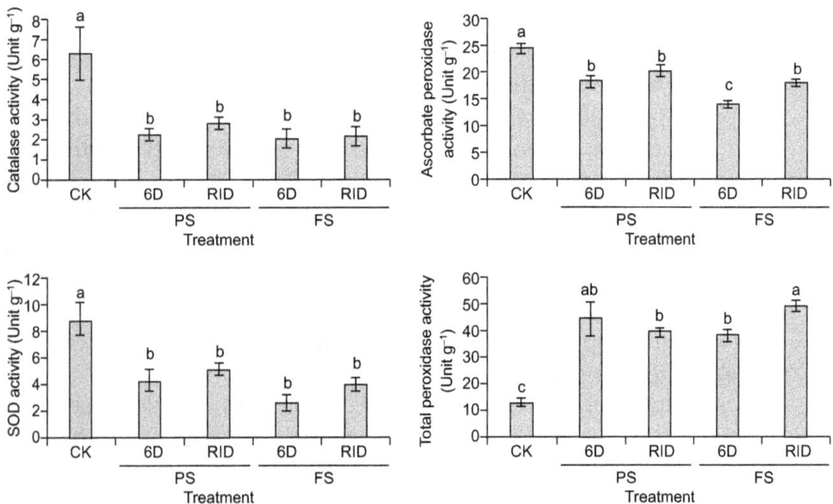

Fig. 5.20: Redox enzymes in rice under control, partial submergence and full submergence.

Total RNAs were isolated from shoots of 14-day-old seedlings after 24 and 48 hours of submergence treatment and levels of Rboh A–I mRNA were determined. Relative amounts of transcripts were calculated and normalized to that of ubiquitin mRNA. Values represent means standard deviation from five biologically independent experiments. Values with different letters are significantly different at $P < 0.05$, according to post-hoc LSD (Fig. 5.21).

Lipid peroxidation produces malondialdehyde (MDA). The amount of MDA indicates the extent of damage resulted by ROS in the tissue (Damanik et al., 2012). This is formed due to oxidative stress which induces membrane damage. Significant negative co-relation has been found between formation of MDA with ascorbate concentration and the survival rate in rice plants (Kawano et al., 2002). Damanik et al. (2012) indicated the significant difference between MDA content in rice seedlings under anoxic and aerobic conditions, and the extended duration of anaerobic conditions increased the lipid peroxidation. MDA content is significantly higher in sensitive cultivars during flooding and after exposure to air (Panda and Sarkar, 2013).

Sub1A restricted the accumulation of ROS and diminished oxidative damage during submergence stress, through the accumulation of negative regulators of GA signalling, the DELLA protein SLR1, and the SLR-like 1 (Fukao et al., 2006).

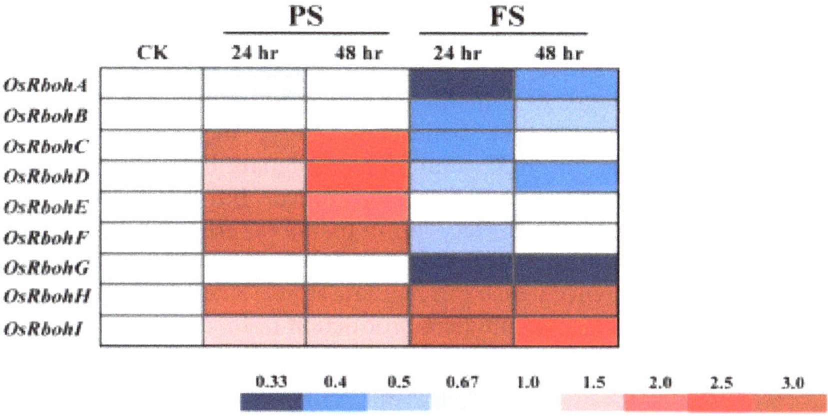

Fig. 5.21: OsRboh activity in rice under partial and full submergence.

5.5 De-submergence and Recovery

Abiotic stresses in plants are often transient, and the recovery phase following stress removal is critical. Flooding, a major abiotic stress that negatively impacts plant biodiversity and agriculture, is a sequential stress where tolerance is strongly dependent on viability underwater and during the post-flooding period. It was shown that in *Arabidopsis thaliana* accessions (Bay-0 and Lp2-6), different rates of submergence recovery correlate with submergence tolerance and fecundity. A genome-wide assessment of ribosome-associated transcripts in Bay-0 and Lp2-6 revealed signalling network-regulating recovery processes. Differential recovery between the accessions was related to the activity of three genes: RESPIRATORY BURST OXIDASE HOMOLOG D, SENESCENCE-ASSOCIATED GENE113, and ORESARA1, which function in a regulatory network involving a reactive oxygen species (ROS) burst upon de-submergence and the hormones: abscisic acid and ethylene. This regulatory module controls ROS homeostasis, stomatal aperture, and chlorophyll degradation during submergence recovery. This work uncovers a signalling network that regulates recovery processes following flooding to hasten the return to pre-stress homeostasis.

5.6 Signalling Network-mediating Post-submergence Recovery

Following prolonged submergence, the shift to a normoxic environment generates the post-submergence signals ROS, ethylene, and ABA. A ROS burst upon re-oxygenation occurs due to reduced scavenging and increased production in Bay-0 from several sources, including RBOHD activity. While excessive ROS accumulation is detrimental and can cause cellular damage, ROS-mediated signalling is required to trigger downstream processes that benefit recovery, including enhanced antioxidant capacity for ROS homeostasis. Signals triggering RBOHD induction following de-submergence are unclear, but hormonal control is most likely to be involved. Recovering plants experience physiological drought due to reduced root conductance, resulting in increased ABA levels post-submergence which can regulate stomatal movements to offset excessive water loss. High ethylene production in Bay-0 caused by ACC oxidation upon re-aeration can counter drought-induced stomatal closure via induction of the protein phosphatase 2C SAG113, accelerating water loss and senescence. Higher transcript abundance of SAG113 in Bay-0 is also positively regulated by ABA, and could be a means to speed up water loss and senescence in older leaves. Ethylene also accelerates chlorophyll breakdown via the NAC transcription factor ORE1. The timing of stomatal reopening during recovery is critical for balancing water loss with CO_2 assimilation, and is likely to be regulated by post-submergence ethylene-ABA dynamics and signalling interactions.

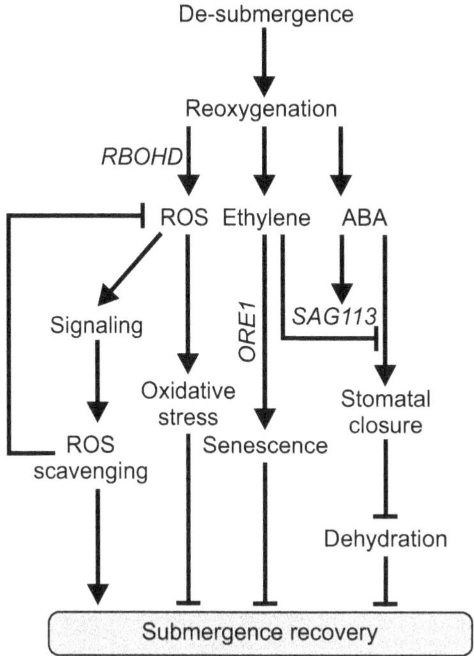

Strategies of adaptation to excess water stresses in the form of submergence or waterlogging in rice plants

Rice can adapt to submergence by internal aeration and growth control. For internal aeration, rice develops longitudinally, forming aerenchyma and leaf gas films. On the other hand, some rice cultivars can survive under submergence by using special strategies of growth control: a quiescence strategy or an escape strategy. The Submergence-1A (Sub1A) gene is responsible for the quiescence strategy, which is important for survival under flash-flood conditions. The SNORKEL1 (SK1) and SNORKEL2 (SK2) genes are responsible for the escape strategy, which is important for survival under deepwater-flood conditions. Rice can adapt to soil waterlogging by forming aerenchyma and a barrier to radial O_2 loss (ROL) in the roots (Fig. 5.22).

5.7 External and Internal Stresses Induced during Submergence and following De-submergence in Plants

When immersed in water, plants encounter drastic changes in environmental parameters (external stresses), triggering a variety of

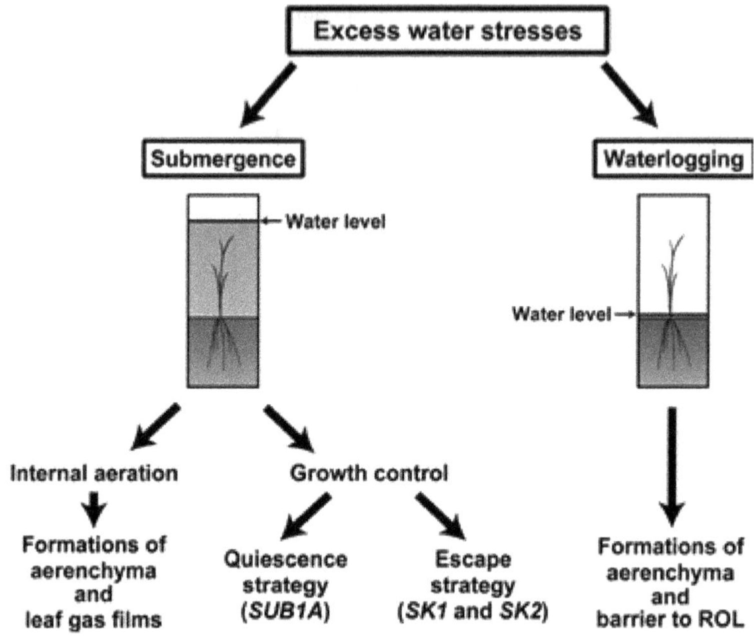

Fig. 5.22: Strategies adapted for excess water stresses in rice.

internal stresses. When floodwaters recede, submerged plants are suddenly exposed to aerobic conditions, inducing additional external and internal challenges. To overcome submergence and post-submergence stresses, plants require tolerance to multiple stresses that occur simultaneously or sequentially over a flood event (Fig. 5.23).

It was clarified earlier that leaf water potential, stomatal aperture, and photosynthetic rate in rice plants decreases with increase in solar radiation and vapour pressure deficit, even though rice plants grew where sufficient water was supplied to their roots in submerged paddy field, and that the decrease in leaf water potential, stomatal aperture, and photosynthetic rate were more remarkable in rice plants with low root activity or with poor root system. The present study was conducted to investigate the relationship between water uptake and transpiration rates and the effects of this relationship on leaf water potential and stomatal aperture through their diurnal changes, and to discuss the characteristics for maintaining water balance in rice plants. Transpiration rate was higher than water uptake rate in the morning when transpiration increases rapidly with rapid increase in solar radiation and vapour pressure deficit. Both rates were practically the same at midday and the transpiration rate was lower than water uptake in the evening when transpiration rate decreased rapidly with rapid decrease in solar radiation and vapour pressure deficit (Fig. 5.24A). The difference between water uptake and transpiration

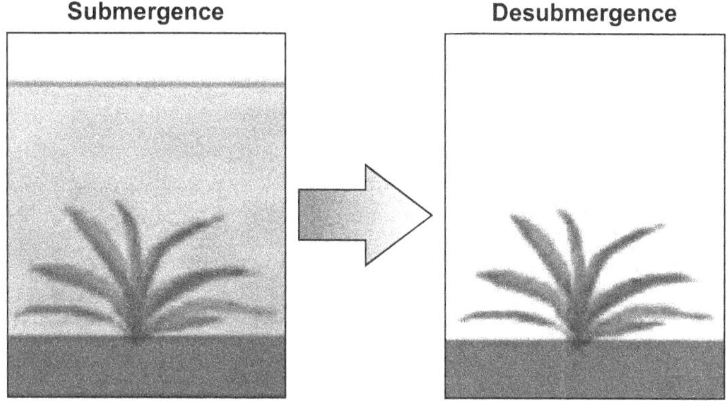

External stress
- Slow gas dithision
- Low light
- High risk of infection
- Salinity (coastal areas)

Internal stress
- Oxygen deprivation
- Limited transpiration
- Stavation
- Nutrient deficiency
- Accumulation of toxic end-products
- Osmotic stress and salt toxicity (coastal areas)

External stress
- Higher oxygen concentration
- Higher light
- High risk of infection
- High risk of insect attack
- Soil nutrient leaching
- Salinity (coastal areas)

Internal stress
- Oxidative stress
- Dehydration
- Photoinhibition
- Nutrient deficiency
- Osmotic siress and salttoxicity (coastal areas)

Fig. 5.23: Comparative facts of submergence and de-submergence.

rates was very small even when transpiration rate was changing rapidly (Fig. 24B). In case of reduced water uptake due to low water potential of culture solution or NaN_3 treatment to roots, transpiration rate decreased remarkably due to increase in stomatal closure during daytime. Therefore, the difference between water uptake and transpiration rates did not increase so much and leaf water potential decreased very little even at midday with high transpiration demand compared with decrease of water uptake rate and stomatal aperture. These results suggested that water balance in rice plants was maintained by the process as follows: There is too much transpiration during daytime due to high solar radiation and vapour pressure deficit, so water uptake could not overtake transpiration, and leaf water potential decreased to a certain extent. As stomata in rice plants were very sensitive to change in leaf water potential compared with those of other plants, stomata closed very rapidly in response to decrease of water potential, so that transpiration rate decreased to almost the same rate as water uptake. Therefore, the difference between water uptake and transpiration rates was very small, and decrease of leaf water potential was prevented. Furthermore, in case of rice plants with reduced water uptake, due to low water potential of culture solution or NaN_3 treatment to roots, all were the same as in the process of maintaining water balance. From these results and the high correlation between stomatal aperture and photosynthetic rate, it was assumed that water uptake ability directly affected photosynthetic rate under sufficient solar radiation in rice plants with stomata responding very sensitively to change of leaf water potential. Further, it was suggested that rapid wilting often observed in rice, soybean and cucumber under very large vapour pressure deficit or on water-saturated soil in the rainy season, both, could arise from decrease of water uptake ability and loss of sensitivity of stomata to decrease in leaf water potential.

Caryopses of floating rice (*Oryza sativa* L. cv. Habiganj Aman II) were germinated and seedlings grown as described previously (Azuma et al., 1990a). Stem segments, 20-cm long and containing the youngest elongating internodes, were prepared from two- to three-month-old plants by the method of Raskin and Kende (1984a). Stems were excised under water to avoid entry of air into the xylem. Segments in which the initial lengths of the youngest internodes ranged from 2 to 6 cm were used for the experiments.

Little is known about the promotive effect of ethylene on transpiration, although it has been reported that ethylene affects stomatal closure in some plant species but not in others (Abeles et al., 1992). It is well established that exposure of plants to water-deficit stress results in accumulation of ABA in leaves of various species (Mansfield and McAinsh, 1995). Also, ABA is known to inhibit stomatal opening. In floating rice, it showed previously that ethylene greatly decreases the levels of ABA in shoots even under low RH conditions (Azuma et al., 1995). Therefore, the decrease of ABA, induced by ethylene, might cause the promotion of transpiration in stem segments through stomatal opening (Fig. 5.25).

Floating or deepwater rice plants show markedly high rates of internodal elongation under submerged conditions (Catling, 1992). This response to submergence is known to be mediated by an interaction between ethylene, gibberellin (GA), and abscisic acid (ABA; *see* review by Kende et al., 1998). When applied to stem sections of floating rice, ethylene and GA promote internodal elongation (Raskin and Kende, 1984b; Suge, 1985; Azuma et al., 1990b), whereas ABA reduces the rate of internodal elongation induced by either ethylene or GA (Hoffmann-Benning and Kende, 1992; Azuma et al., 1990b, 1995). It was found previously that ethylene-induced internodal elongation is inhibited at low relative humidity (RH), while GA-induced internodal elongation is barely affected by humidity (Azuma et al., 1991). This difference with RH could not be explained in terms of changes in endogenous levels of ABA because

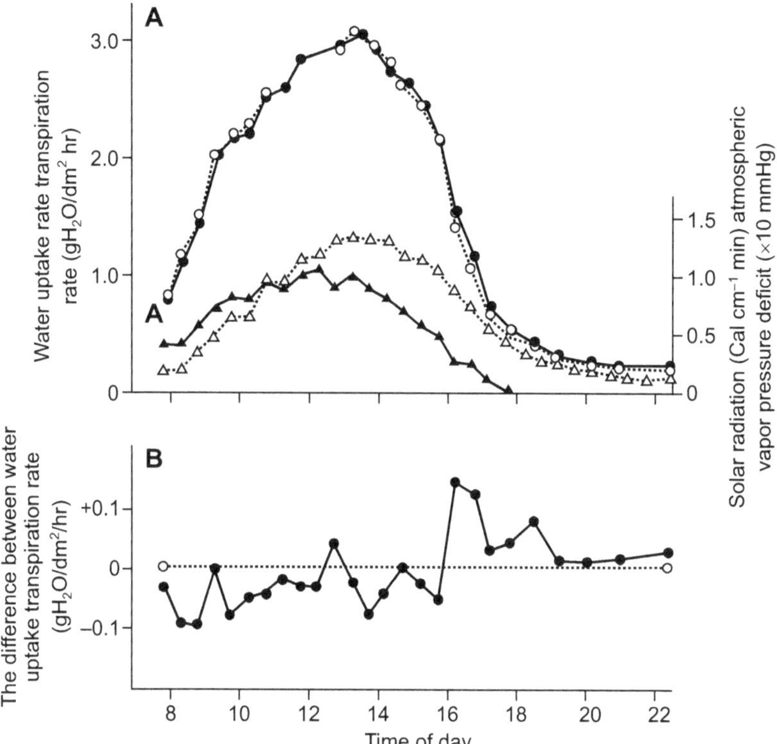

Fig. 5.24: Water uptake rate/transpiration rate (A) and the difference (B).

ethylene decreased the level of ABA in internodes at any RH (Azuma et al., 1995).

In general, cell elongation is driven by uptake of water into the cell. Flow of water into the cell is dependent on the difference in water potential between the soluble phase of the cell and the apoplast around the cell. The water potential of the soluble phase of the apoplast must be influenced by the humidity of the surroundings because of changes in the rate of transpiration. Therefore, there is a possibility that the difference in the response of floating rice internodes to ethylene and GA at low RH is due to the difference in the rate of transpiration. In the present study, researchers investigated the effects of ethylene and GA on transpiration from floating rice stems at various RHs.

Regardless of the temperature, duration, and depth of the flood, flood has a major impact and threatens plants with a shortage of carbohydrates and cellular energy, which hinders growth and development (Fig. 5.26). Rain-water flooding is usually clear water, which results in less crop damage as compared to turbid or silted water. Hence, comprehensive knowledge of the relationship between plant survival and flood water

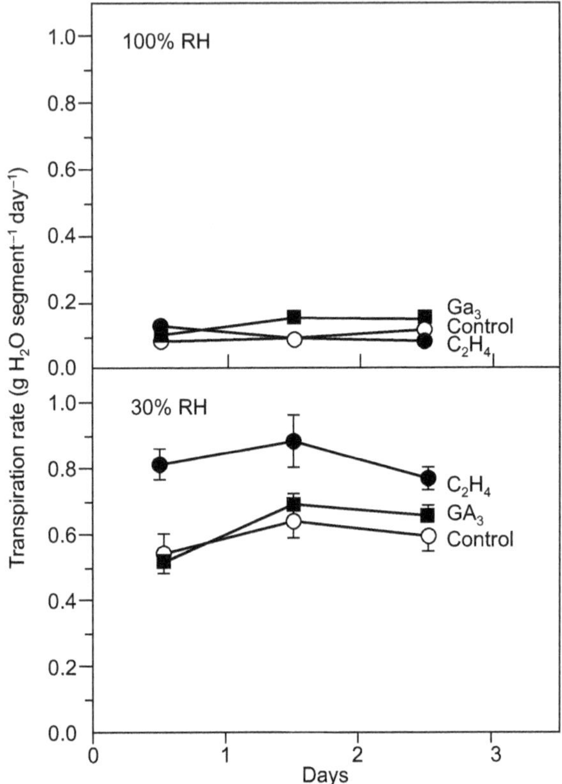

Fig. 5.25: Changes with time in the rate of transpiration of stem segments treated with ethylene or GA_3 at different rates of humidity for growth conditions.

qualities is a prerequisite for the development of flood tolerance and sustainable flood-management practices.

Submergence causes changes in gene expression, which coordinates morphological and metabolic adaptations to stress (Bailey-Serres and Voesenek, 2008). In rice-growing areas, crops frequently face floods which may be a flash flood resulting in complete submergence. Transcription factors, such as group VII ethylene response factors (ERF-VIIs) regulate the expression of a varied range of genes involved in adaptive responses to excess moisture or low oxygen (Voesenek and Bailey-Serres, 2015). In response to submergence, deepwater paddy enhances cell division and elongation growth by ethylene-mediated pathway through activation of ACC synthase gene. Various genes, like SNORKEL1 and SNORKEL2 (ERF VII) were also reported in deepwater rice (Hattori et al., 2009) that encode two ethylene-responsive factors (ERF) DNA-binding proteins. In lowland rain-fed paddy, escape and quiescence strategies are followed

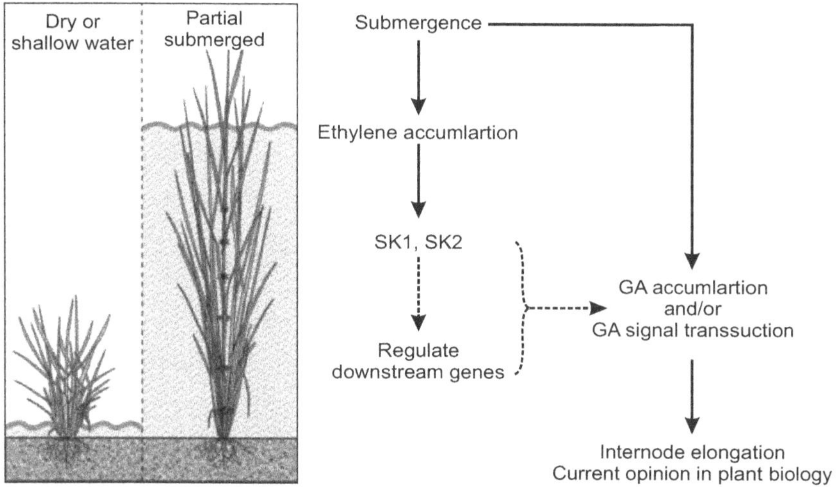

Fig. 5.26: Mechanism in partial submergence (waterlogged) rice.

to get rid of hypoxia conditions due to flooding. In rice, an ERF-VII TF gene (Sub1A) is known as the main regulator of submergence tolerance, permitting plants to withstand complete submergence for 14–16 days (Fukao et al., 2006; Xu et al., 2006). It is demonstrated that Sub1A enhances inhibition of gibberellic acid (GA) signalling by up-regulation of rice DELLA proteins, like SLR1 (Slender Rice 1) and SLRL1 (SLR1-like1), known as negative regulators of GA (Fukao and Bailey-Serres, 2008). Consequently, GA signalling is inhibited, causing limited elongation of growth (Fig. 5.27).

Low oxygen also regulates the function of various K^+ channels in mammals (Wang et al., 2017). Under waterlogging, K^+ concentration in the soil considerably drops, and the uptake of K^+ by plants gets restricted due to reduced hydraulic conductivity. Further, this suggests that low oxygen and flooding can increase internal K^+ concentration in specific tissues and cell types due to altered ion channel activities. A recent study noted that under combined oxygen-deprived and K^+ sufficient conditions, there was accumulation of AtRAP2.12 protein (an Arabidopsis ERF-VII) improved by a Raf-like mitogen-activated protein kinase kinase kinase (MAPKKK) and HCR1 (Shahzad et al., 2016).

Greater genotypic variability was observed for plant height, elongation, and survival percentage, absolute growth rate, non-structural carbohydrate retention capacity, chlorophyll content, different chlorophyll fluorescence parameters (FPs) characteristics, and re-generation growth at re-emergence. Twenty days of submergence caused greater damage even

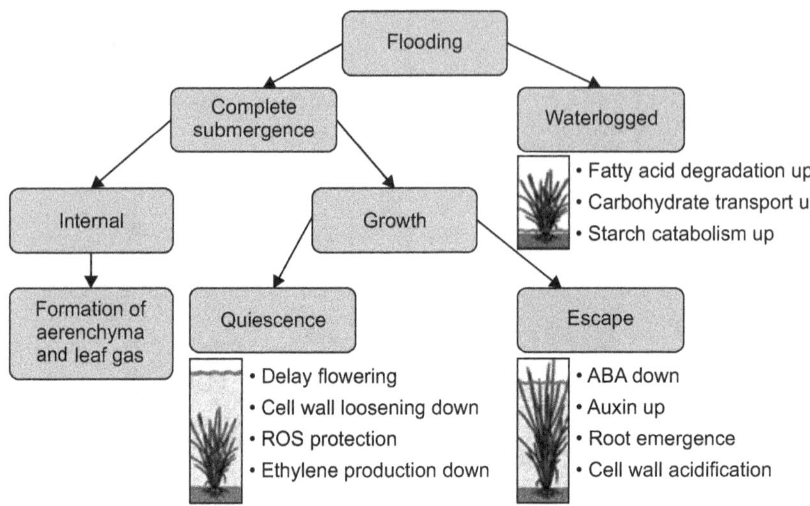

Fig. 5.27: Functions of different types of flooding.

in submergence 1 (Sub1) introgressed cultivars compared to the 14 days of submergence. The FPs, carbohydrate content, and dry weight at the end of submergence showed positive and highly significant association with regeneration growth. The presence of Sub1-associated primers, either SC3 or ART5, was noticed even in greater elongating types of rice genotypes. These genotypes possess one or more of the adaptive traits required for the flood-prone ecosystem and ranged from temporary submergence of one to two weeks to long periods of stagnant water tolerance.

References

Abeles, F.B., Morgan, P.W., and Saltveit Jr., M.E. (1992). *Ethylene in Plant Biology*, Academic Press, San Diego, CA.

Airaki, M., Leterrier, M., Mateos, R.M., Valderrama, R., Chaki, M., Barroso, J.B. et al. (2012). Metabolism of reactive oxygen species and reactive nitrogen species in pepper (*Capsicum annuum* L.) plants under low temperature stress. *Plant, Cell and Environment*, 35: 281-295.

Angaji, S.A., Septiningsih, E.M., Mackill, D.J., and Ismail, A.M. (2010). QTLs associated with tolerance of flooding during germination in rice (*Oryza sativa* L.). *Euphytica*, 172: 159–168.

Armstrong, W. (1979). Aeration in higher plants. *Advances in Botanical Research*, 7: 225–332.

Azuma, T., Mihara, F., Uchida, N., Yasuda, T., and Yamaguchi, T. (1991). Influence of humidity on ethylene-induced internodal elongation in floating rice. *Plant Cell Physiol.*, 32. Doi:10.1093/oxfordjournals.pcp.a078080.

Azuma, T., Uchida, N., Yasuda, T., and Yamaguchi, T. (1994). Gibberellin induced ethylene production in internodes of floating rice. *Jpn. J. Trop. Agr.*, 38: 78–82.

Azuma, T., Hirano, T., Deki, Y., Uchida, N., Yasuda, T., and Yamaguchi, T. (1995). Involvement of the decrease in levels of abscisic acid in the internodal elongation of submerged floating rice. *J. Plant Physiol.*, 146: 323–328.

Azuma, T., Hatanaka, T., Uchida, N., and Yasuda, T. (2003). Interactions between abscisic acid, ethylene and gibberellins in intermodal elongation in floating rice: The promotive effect of abscisic acid at low humidity. *Plant Growth Regul.*, 41: 105–109.

Bailey-Serres, J., and Voesenek, L.A.C.J. (2008). Flooding stress: Acclimations and genetic diversity. *Annu. Rev. Plant Biol.*, 59: 313–339.

Bailey-Serres, J., Fukao, T., Ronald, P.C., Ismail, A.M., Heuer, S., and Mackill, D.J. (2010). Submergence tolerant rice: Sub1's journey from landrace to modern cultivar. *Rice*, 3: 138–147.

Burg, S.P., and Burg, E.A. (1965). Gas exchange in fruits. *Physiol. Plant*, 18. Doi. org/10.1111/j.1399-3054.1965.tb06946.x.

Catling, D. (1992). *Rice in Deep Water*. MacMillan Press, London.

Chakraborty, K., Mondal, S., Ray, S., Samal, P., Pradhan, B., Chattopadhyay, K. et al. (2020). Tissue tolerance coupled with ionic discrimination can potentially minimize the energy cost of salinity tolerance in rice. *Front. Plant Sci.*, 11: 265.

Colmer, T.D., Armstrong, W., Greenway, H., Ismail, A.M., Kirk, G.J.D., and Atwell, B.J. (2014). Physiological mechanisms in flooding tolerance of rice: Transient complete submergence and prolonged standing water. *Progress in Botany*, 75: 255–307.

Colmer, T.D., and Voesenek, L.A.C.J. (2009). Flooding tolerance: Suites of plant traits in variable environments. *Functional Plant Biology*, 36: 665–681.

Colmer, T.D., and Pedersen, O. (2008). Oxygen dynamics in submerged rice (*Oryza sativa*). *New Phytol.*, 178: 326–334.

Colmer, T.D., and Pedersen, O. (2008). Underwater photosynthesis and respiration in leaves of submerged wetlandplants:gas films improve co2 and O2 exchange. *New Phytol.*, 117: 918–926.

Damanik, R.I., Ismail, M.R., Shamsuddin, Z., Othman, S., Zain, A.M., and Maziah, M. (2012). Response of antioxidant systems in oxygen deprived suspension cultures of rice (*Oryza sativa* L.). *Plant Growth Regul.*, 67(1): 83–92.

Das, K.K., and Sarkar, R.K. (2001). Post flood changes on the status of chlorophyll, carbohydrate and nitrogen content and its association with submergence tolerance in rice. *Plant Archive*, 1: 15-19.

Das, K., and Roychoudhury, A. (2014). Reactive oxygen species (ROS) and response of antioxidants as ROS-scavengers during environmental stress in plants. *Front. Environ. Sci.*, 2: 53.

Dauphinee, A., Warner, T., and Gunawardena, A. (2014). A comparison of induced and developmental cell death morphologies in lace plant (*Aponogeton madagascariensis*) leaves. *BMC Plant Biology*, 14: 1589.

Fukao, T., Xu, K., Ronald, P.C., and Bailey-Serres, J. (2006). A variable cluster of ethylene response factor likegenes regulates metabolic and developmental acclimation responses to submergence in rice. *Plant Cell*, 18: 2021–2034.

Fukao, T., and Bailey-Serres, J. (2008). Submergence tolerance conferred by Sub1A is mediated by SLR1 and SLRL1 restriction of gibberellin responses in rice. *Proc. Natl. Acad. Sci.*, USA, 105(43): 16814–16819.

Fukao, T., Yeung, E., and Bailey-Serres, J. (2011). The submergence tolerance regulator SUB1A mediates crosstalk between submergence and drought tolerance in rice. *Plant Cell*, 23: 412– 427.

Gill, S.S., and Tuteja, N. (2010). Reactive Oxygen Species and Antioxidant Machinery in Abiotic Stress Tolerance in Crop Plants. *Plant Physiology and Biochemistry*, 48: 909–930.

Hattori, Y., Nagai, K., Furukawa, S., Song, X.J., Kawano, R., Sakakibara, H. et al. (2009). The ethylene response factors SNORKEL1 and SNORKEL2 allow rice to adapt to deep water, *Nature*, 460(7258): 1026–U1116.

Hoffmann-Benning, S., and Kende, H. (1992). On the role of abscisic acid and gibberellin in the regulation of growth in rice. *Plant Physiol.*, 99: 1156–1161.

Houmani, H., Rodridguez-Ruiz, M., and Palma, J.M. (2018). Mechanical wounding promotes local and long distance response in the halophyte *Cakile maritima* through the involvement of the ROS and RNS metabolism. *Nitric Oxide*, 74: 93–101.

Huang, Y.-M. (2019). Examining students continued use of desktop services – Prospectives from expectation-confirmation and social influence. *Computers in Human Behaviour*, 96: 23–31.

Ismail, A.M., Ella, E.S., Vergara, G.V., and Mackill, D.J. (2009). Mechanisms associated with tolerance to flooding during germination and early seedling growth in rice (*Oryza sativa*). *Ann. Bot.*, 103: 197–209.

Jackson, M.B., and Ram, P.C. (2003). Physiological and Molecular Basis of Susceptibility and Tolerance of Rice Plants to Complete Submergence. *Annals of Botany*, 91: 227–241.

Jain, M., Tyagi, A.K., and Khurana, J.P. (2006). Molecular characterization and differential expression of cytokinin-responsive type-A response regulators in rice (*Oryza sativa*). *BMC Plant Biol.*, 6: 1.

Joshi, R., Vig, A.P., and Singh, J. (2013). Vermicompost as soil supplement to enhance growth, yield and quality of *Triticum aestivum* L.: A field study. *Int. J. Recycl. Org. Waste Agric.*, 2(1): 16.

Jung, K.H., Seo, Y.-S., Walia, H., Cao, P., Fukao, T., Canlas, P.E. et al. (2010). The submergence tolerance regulator *Sub1A* mediates stress-responsive expression of *AP2/ERF* transcription factors. *Plant Physiol.*, 152: 1674–1692.

Kaur, G., Sharma, A., Guruprasad, K., and Pati, P.K. (2014). Versatile roles of plant NADPH oxidase and emerging concepts. *Biotechnology Advances*, 32: 551–563.

Kawano, N., Ella, E., Ito, O., Yamauchi, Y., and Tanaka, K. (2002). Metabolic changes in rice seedlings with different submergence tolerance after de-submergence. *Environ. Exp. Bot.*, 47(3): 195–203.

Kende, H., van der Knaap, E., and Cho, H.T. (1998). Deepwater Rice: A model plant to study stem elongation. *Plant Physiology*, 118: 1105–1110.

Kepinski, S. (2006). Integrating hormone signalling and patterning mechanisms in plant development. *Curr. Opin. Plant Biol.*, 9: 28–34.

Konings, H., and Jackson, M.B. (1979). A relationship between rates of ethylene production by roots and the promoting or inhibiting effects of exogenous ethylene and water on root elongation. *Zeitschrift fu¨r Pflanzenphysiologie*, 92: 385–397.

Ku, H.S., Suge, H., Rappaport, L., and Pratt, H.K. (1970). Stimulation of rice coleoptile growth by ethylene. *Planta*, 90: 333–339.

Magneschi, L., and Perata, P. (2009). Rice germination and seedling growth in the absence of oxygen. *Ann. Bot.*, 103(2): 181–196.

Mansfield, T.A., and McAinsh, M.R. (1995). Hormones as regulators of water balance. pp. 598–616. *In*: Davies, PfJf (Ed.). *Plant Hormones*, Kluwer, Dordrecht.

Miro, B., Longkumer, T., Entila, F.D., Kohli, A., Ismail, A.M. (Oct. 2017). Rice seed germination underwater: morpho-physiological responses and the bases of differential expression of alcoholic fermentation enzymes. *Front. Plant Sci.*, 8: 1857.

Nabi, R.B.S., Tayade, R., Hussain, A., Kulkarni, K.P., Imran, Q.M., Mun, B.G. et al. (2019). Nitric oxide regulates plant responses to drought, salinity, and heavy metal stress. *Environmental and Experimental Botany*, 161: 120–133.

Panda, D., and Sarkar, R.K. (2013). Characterization of leaf gas exchange and anti-oxidant defence of rice (*Oryza sativa* L.) cultivars differing in submergence tolerance owing to complete submergence and consequent re-aeration. *Agric. Res.*, 2(4): 301–308.

Pedersen, O., Rich, S.M., and Colmer, T.D. (2009). Surviving floods:gas films improve O2and CO2 exchange, root aeration and growth of completely submerged rice. *Plant J.*, 58: 147–156.

Raskin, I., and Kende, H. (1983). How does deep water rice solve its aeration problem. *Plant Physiol.*, 72: 447–454.

Raskin, I., and Kende, H. (1984). Role of gibberellin in the growth response of submerged deep-water rice. *Plant Physiol.*, 76: 947–950.

Raskin, I., and Kende, H. (1984a). Regulation of growth in stem sections of deepwater rice. *Planta*, 160: 66–72.

Raskin, I., and Kende, H. (1984b). Role of gibberellin in the growth response of deep water rice. *Plant Physiol.*, 76: 947–950.

Sakagami, J., Joho, Y., and Ito, O. (2009). Constracting physiological responses by cultivars *Oryza sativa* and *O. glaberrima* to prolonged submergence. *Ann. Bot.*, 103: 171–180.

Sarkar, P., and Gladish, D.K. (2012). Hypoxic stress triggers a programmed cell death pathway to induce vascular cavity formation in *Pisum sativum* roots. *Physiol. Plant*, 146: 413– 426.

Sarkar, R.K. (1998). Saccharide content and growth parameters in relation with flooding tolerance in rice. *Biol. Plantarum,*, 40: 597–603.

Sarkar, R.K., and Bhattacharjee, B. (2011). Rice Genotypes with SUB1 QTL Differ in Submergence Tolerance, Elongation Ability during Submergence and Re-submergence and Regeneration Growth at Re-emergence. *Rice*, 5: 7.

Setter, T.L., Kupkanchanakul, T., Pakkinaka, l., Aguru, Y., and Greenway, H. (1987b). Mineral nutrients in flood-water and floating rice growing in water depths up to two metres. *Plant Soil*, 104: 147–150.

Setter, T.L., Waters, I., Wallace, I., Bhekasut, P., and Greenway, H. (1989). Submergence of rice. I. Growth and photosynthetic response to CO_2 enrichment of floodwater. *Aust. J. Plant Physiol.*, 16: 251–263.

Shahzad, Z., Canut, M., Tournaire-Roux, C., Martinière, A., Boursiac, Y., Loudet, O. et al. (Sept. 2016). Potassium-dependent Oxygen Sensing Pathway Regulates Plant Root Hydraulics. *Cell*, 167(1): 87–98.e14.

Singh, S., Mackill, D.J., and Ismail, A.M. (2014). Physiological basis of tolerance to complete submergence in rice involves genetic factors in addition to the SUB1 gene. *AoB Plants*, 6: plu060.

Stünzi, J.T., and Kende, H. (1989). Light-dependent short-term modulations of elongation in rice plants. *Plant Cell Physiol.*, 30: 415–422.

Suge, H. (1985). Ethylene and gibberellin: regulation of internodal elongation and nodal root development in floating rice. *Plant Cell Physiol.*, 26: 607–614.

Verboven, P., Pedersen, O., Ho, Q.T., Nicolai, B.M., and Colmer, T.D. (2014). The mechanism of improved aeration due to gas films on leaves of submerged rice. *Plant, Cell & Environment*, 37: 243–252.

Voesenek, L.A.C.J., Benschop, J.J., Boui Torrent, J., Cox, M.C.H., Groeneveld, H.W., Millenaar, F.F. et al. (2003). Interactions between plant hormones regulate submergence-induced shoot elongation in the flooding tolerant dicot *Rumex palustris*. *Annals of Botany*, 91: 205–211.

Voesenek, L.A.C.J., and Bailey-Serrres, J. (2015). Flood adaptive traits and processes:an overview. *New Phytol.*, 206: 57–73.

Wang, M., Lu, Y., Botella, J.R., Mao, Y., Hua, K., and Zhu, J.K. (July 2017). Gene targeting by homology-directed repair in rice using a geminivirus-based CRISPR/Cas9 System, *Mol. Plant.*, 10(7): 1007–1010. Doi: 10.1016/j.molp.2017.03.002. Epub 2017 Mar. 16. PMID: 28315751.

Waters, I., Armstrong, W., Thomson, C.J., Setter, T.L., Adkins, S., Gibbs, J. et al. (1989). Diurnal changes in radial oxygen loss and ethanol metabolism in roots of submerged and non-submerged rice seedlings. *New Phytol.*, 113: 479–491.

Wilkins, O., Bräutigam, K., and Campbell, M.M. (2011). Time of day shapes *Arabidopsis* drought transcriptomes. *Plant J.*, 63: 715–727.

Winkel, A., Colmer, T.D., Ismail, A.M., and Pedersen, O. (2013). Internal aeration of paddy field rice (*Oryza sativa*) during complete submergence—Importance of light and floodwater O_2. *New Phytol.*, 197: 1193–1203.

Xu, K., Xu, X., Fukao, T., Canlas, P., Maghirang-Rodriguez, R., Heuer, S. et al. (Aug. 2006). Sub1A is an ethylene-response-factor-like gene that confers submergence tolerance to rice, *Nature*, 0:442(7103): 705–8.

Yamauchi, T., Fukazawa, A., and Nakazono, M. (2017 and 2017a). Metallothjoneingenes encoding ROS scavenging enzymes are down-regulated in the root cortex during inducible aerenchyma formation in rice. *Plant Signal Behav.*, 12: e1388976.

6

Molecular and Genetic Aspects

Flooding or submergence is one of the major environmental stressors affecting many man-made and natural ecosystems worldwide. The increase in the frequency and duration of heavy rainfall due to climate change has negatively affected plant growth and development, which eventually causes the death of plants if it persists for days. Most crops, especially rice, being a semi-aquatic plant, are greatly affected by flooding, leading to yield losses each year. Genetic variability in the plant response to flooding includes the quiescence scheme, which allows underwater endurance for a prolonged period, escape strategy through stem elongation, and alterations in plant architecture and metabolism. Investigating the mechanism for flooding survival in wild species and modern rice has yielded significant insight into developmental, physiological, and molecular strategies for submergence and waterlogging survival. Significant progress in the breeding of submergence-tolerant rice varieties has been made during the last decade following the successful identification and mapping of a quantitative trait locus for submergence tolerance, designated as SUBMERGENCE 1 (Sub1) from the FR13A landrace. Using marker-assisted backcrossing, the Sub1 QTL (quantitative trait locus) has been incorporated into many elite varieties within a short time and with high precision as compared with conventional breeding methods. Despite the advancement in submergence tolerance, for future studies, there is a need for practical approaches to explore genome-wide association studies (GWA) and QTL in combination with specific tolerance traits, such as drought, salinity, disease and insect resistance.

Evolutionary analyses focusing on Sub1A-1 and flash-flood tolerance in wild rice have been performed by sequencing Sub1A-related genes. Sub1A-1 alleles are found in some accessions of wild rice (*O. nivara* and *O. rufipogon*, which belong to the AA genome group) together with *O. sativa* (Fukao et al., 2009; Niroula et al., 2012; Zhao et al., 2018). The geographical distribution of these accessions harbouring Sub1A-1 suggests that Sub1A-1 from wild species might have introgressed around the Ganges Basin and subsequently spread to other areas of South Asia (Pucciariello and Perata, 2013). By contrast, Sub1A-1 is absent in submergence-tolerant accessions of wild rice with the CC genome (*O. rhizomatis* and *O. eichingeri*) and CCDD genome (*O. grandiglumis*) (Niroula et al., 2012; Okishio et al., 2015) suggesting the presence of a Sub1A-independent mechanism in these rice species. Future elucidation of the Sub1A-independent mechanism may contribute to future breeding of cultivated rice with strong flash-flood tolerance.

Flooding inundates approximately 20% of lowland rice paddies throughout the world each year, leading to significant yield losses in Asia and the US. Consequentially, a major goal of rice breeders is to generate cultivars that can withstand 14 days or more of submergence at the seedling stage. In the US, farmers typically plant rice by aerial broadcast of pre-germinated rice seeds into shallow paddies. This pre-germination method requires additional seed use to ensure good stand establishment; however the shallowness of paddies does not adequately discourage weeds. The development of cultivars that can be dry-seeded into flooded fields will significantly reduce the levels of herbicide applied at the pre- and post-emergence stages. This project investigates the molecular basis of submergence tolerance by elucidating the role of the Sub1A gene of the complex Sub1 locus in the acclimation response to this stress. Thus far the project has determined that Sub1A is indeed a DNA-binding protein.

The traditional varieties and landraces of rice possess variable levels of tolerance to submergence stress, but gene discovery and utilization of these resources has been limited to the Sub1A-1 allele from variety FR13A. Therefore, researchers analysed the allelic sequence variation in three Sub1 genes in a panel of 179 rice genotypes and its association with submergence tolerance. Population structure and diversity analysis based on a 36-plex genome-wide genic-SNP assay grouped these genotypes into two major categories representing Indica and Japonica cultivar groups with further sub-groupings into Indica, Aus, Deepwater and Aromatic Japonica cultivars. Targetted re-sequencing of the Sub1A, Sub1B and Sub1C genes identified 7, 7 and 38 SNPs making 8, 9 and 67 SNP haplotypes, respectively. Haplotype networks and phylogenic analysis revealed evolution of Sub1B and Sub1A genes by tandem duplication and divergence of the ancestral Sub1C gene in that order. The alleles of Sub1 genes in tolerant reference variety FR13A seem to have evolved most recently. However, no consistent association could be found between the Sub1 allelic variation and submergence tolerance probably due to low minor allele frequencies and presence of exceptions to the known Sub1A-1 association in the genotype panel. Researchers identified 18 cultivars with non-Sub1A-1 source of submergence tolerance which, after further mapping and validation in biparental populations, will be useful for development of superior flood tolerant rice cultivars (Singh et al., 2020).

The recent advancements in molecular biology and marker technologies has paved the way for QTL analysis, fine mapping, and subsequent cloning of the submergence-tolerant gene. The molecular basis for submergence tolerance has been well studied for traits linked with flood tolerance in rice. Among these traits, deep-water elongation was the first to capture the attention of plant breeders and geneticists due to its unique quality. Ramaiah and Ramaswami (1940) were the first to reveal that duplicated genes control internode elongation and was designated as ef1 and ef2 (elongation factor), using the conventional breeding method, while studying the inheritance pattern of stem elongation in deep-water rice. Subsequently, Hamamura and Kupkanchankul (1979) reported a partial dominance and five to six genes responsible for the floating of deep-water rice using allele crosses and analyzing the progeny for submergence tolerance. However, according to Tripathi and Rao (1985), a single dominant gene controls early nodal differentiation, whereas Suge (1987) reported the action of complementary genes that control internode elongation in deep-water rice. Eiguchi et al. (1993) concluded that single recessive gene dw3 controlled internode elongation in deep-water rice.

Sub1A belongs to group VII of the AP2/ERF superfamily (ERFVII), whose members are submergence-activated genes that are conserved in angiosperms (Reynoso et al., 2019). Transcriptome analysis using varieties with or without *Sub1A-1* has revealed putative *Sub1A-1*-regulated downstream genes related to anaerobic respiration, hormone responses, and antioxidant systems (Jung et al., 2010; Locke et al., 2018). In addition, several other ERFVII family members are specifically up-regulated under submergence in a rice variety harbouring *Sub1A-1* (Jung et al., 2010). Some members of the ERFVII family have a specific N-terminal motif that functions as an N-degron, targeting the protein for oxygen-dependent degradation; these proteins are stabilized under hypoxia and regulate hypoxia-responsive signalling (Gibbs et al., 2011). While *Sub1A-1* is not under N-end rule regulation, two other ERFVII members, ERF66 and ERF67, are directly transactivated by *Sub1A-1* under submergence (Lin et al., 2019). ERF66 and ERF67 are regulated by the N-end rule pathway and stabilized under hypoxic conditions, suggesting they are responsible for submergence-mediated transcriptional regulation in the quiescent strategy (Lin et al., 2019). Supporting this, overexpression of ERF66 or ERF67 leads to increased expression of anaerobic survival genes (Lin et al., 2019). Further analysis of downstream genes of these ERFVII members will reveal a more detailed mechanism of *Sub1A-1*-mediated regulation as a quiescent strategy.

The major rice quantitative-trait locus *Submergence 1* (*Sub1*) confers tolerance of submergence for about two weeks. To identify novel sources of tolerance, researchers have conducted a germplasm survey with allele-specific markers targeting *Sub1A* and *Sub1C*, two of the three transcription-factor genes within the *Sub1* locus. The objective was to identify tolerant genotypes without the *Sub1A* gene or with the intolerant *Sub1A-2* allele. The survey revealed that all tolerant genotypes possessed the tolerant *Sub1* haplotype (*Sub1A-1/Sub1C-1*), whereas all accessions without the *Sub1A* gene were intolerant. Only the variety James Wee with the *Sub1A-2* allele was moderately tolerant. However, some intolerant genotypes with the *Sub1A-1* allele were identified and RT–PCR analyses were conducted to compare gene expression in tolerant and intolerant accessions. Initial analyses of leaf samples failed to reveal a clear association of *Sub1A* transcript abundance and tolerance. Temporal and spatial gene expression analyses subsequently showed that *Sub1A* expression in nodes and internodes associated best with tolerance across representative genotypes. In James Wee, transcript abundance was high in all tissues, suggesting that some level of tolerance might be conferred by high expression of the *Sub1A-2* allele. To further assess tissue-specific expression, researchers have expressed the GUS reporter gene under the control of the *Sub1A-1* promoter. The data revealed highly specific GUS expression at the base of the leaf sheath and in the leaf collar region. Specific expression in the growing part of rice leaves is well in agreement with the role of *Sub1A* in suppressing leaf elongation under submergence.

6.1 Anaerobic Seed Germination

Direct-seeded rice (DSR), classified as wet DSR, dry DSR, or water DSR, is becoming increasingly popular across the world due to its cost efficiency and convenience (Mahender et al., 2015). Compared with wet and dry DSR, water DSR (whereby seeds are broadcasted in standing water) is more advantageous as it is less labour and time intensive, and restrains weed growth. However, rice (*Oryza sativa* L.) is extremely sensitive to anoxia during germination and early growth of the embryo (Yamauchi et al., 1993; Ismail et al., 2009; Angaji et al., 2010). When using the water DSR method, the rice seeds are completely submerged and suffer hypoxia or anoxia, leading to poor or no germination, seedling death, and poor crop establishment (Yamauchi et al., 1993; Ismail et al., 2009). Anaerobic germination (AG) is the main limiting factor for the large-scale adoption of water DSR. Revealing the biochemical and molecular mechanisms of AG tolerance under flooding conditions and breeding rice varieties with superior AG-tolerant traits could effectively resolve the limitations of the water DSR method.

Sugar starvation caused by low O_2 stress under the submergence condition is the key upstream signal affecting metabolic regulation pathways. The Ca^{2+} signal acts as a secondary messenger to mediate the downstream responses. The calcineurin B-like (CBL) proteins bind to Ca^{2+} and then interact with CIPK15, leading to activation of its kinase activity. Subsequently, the activated CIPK15 physically interacts with SnRK1A, an upstream protein kinase of the transcription factor MYBS1, and activates its activity, consequently elevating the activity of αAmy for seed stored starch degradation. SKIN1/2 and MYBS2 negatively regulate SnRK1A and MYBS1, respectively, thereby repressing αAmy expression. Additionally, seed imbibition also results in the biosynthesis of gibberellin (GA) in the embryo. GA induces the expression of MYBGA in cereal aleurone cells, thereby up-regulating the expression of the αAmy gene. Meanwhile, low O_2-induced OsTPP7 de-suppresses the SnRK1A activity that is inhibited by trehalose 6-phosphate (T6P). OsTPP7 increases the sink strength of the embryo axis–coleoptile by the perception of low sugar availability by enhancing the conversion of T6P to trehalose and leads to a decrease in the T6P/sucrose ratio, thus enhancing starch mobilization for energy production to promote coleoptile elongation. In addition, low O_2 results in the shift of aerobic respiration to anaerobic fermentation, therefore inducing the expression of essential components, including PDC and ADH.

In rice, a total of 10 distinct *αAmy* genes are classified into three subfamilies: subfamilies RAmy1 (A, B, and C), RAmy2A, and RAmy3 (A, B, C, D, E, and F). Most of the *αAmy* genes belong to the RAmy3 subfamily (Huang et al., 1990, 1992), in which only one gene (LOC_Os08g36910) is induced by sugar starvation and O_2 deficiency (Hwang et al., 1999). Thus, RAmy3D is considered an important enzyme in the fermentative metabolism pathway for metabolic shift regulation and in the production of energy in response to submergence stress (Guglielminetti et al., 1995) (Fig. 6.1 and Table 6.1).

To determine the genetic basis of AG in rice, quantitative trait locus (QTL) mapping and genome-wide association study (GWAS) were recently used to identify some QTLs associated with AG. Several studies assigned the elongation of the coleoptile as an indicator trait of the tolerance phenotype in QTL mapping (Jing et al., 2004, 2006; Chen et al., 2012; Wang et al., 2009; Manangkil et al., 2013). Other studies are assigned the seeding survival rate as an indicator of the tolerance phenotype (Angaji, 2008; Angaji et al., 2010; Septiningsih et al., 2013; Baltazar et al., 2014; Kim and Reinke, 2018). Some valuable QTLs were obtained by QTL mapping with the two kinds of indicator. GWAS is a popular and highly efficient strategy for dissecting complex traits, but few reports regarding the excavation of AG tolerance loci via GWAS are available. A GWAS, using 36,901 single nucleotide polymorphisms (SNPs), identified 88 loci

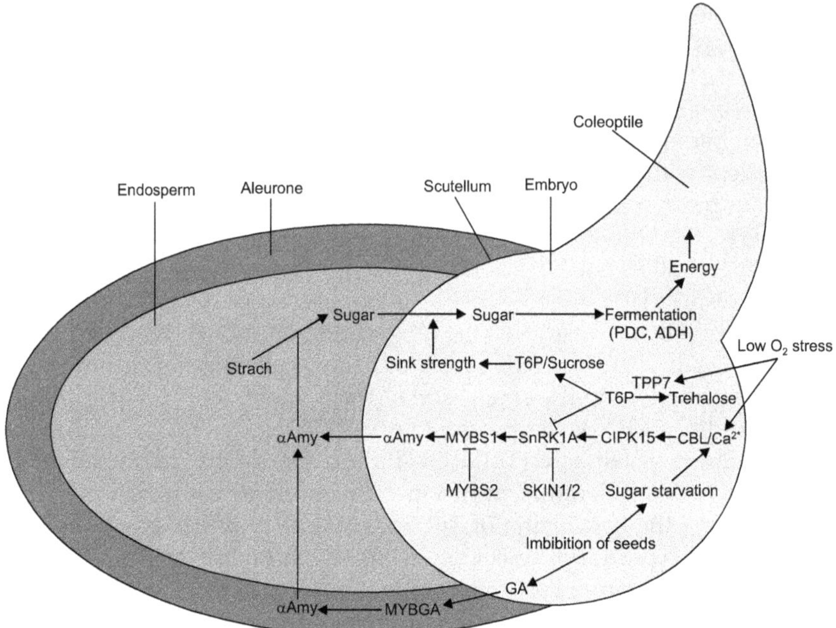

Fig. 6.1: The molecular regulatory pathways and metabolic pathways of rice under hypoxic germination and early seedling growth.

associated with AG tolerance (Hsu and Tung, 2015). Through GWAS of 5291 SNPs in 432 indica varieties, a total of 15 AG tolerance loci were detected by another report (Zhang et al., 2017). Although numerous AG tolerance loci have been identified, only one QTL (qAG-9-2) has been fine mapped and cloned as OsTPP7. Therefore, a large gap between the identification of AG tolerance genes and the breeding of DSR rice varieties still exists. This gap is mainly due to the majority of reported loci being identified based on low-density markers, and few reliable and stable AG tolerance loci have been screened for gene cloning. It is thus imperative that ultra-high density markers are used to evaluate stable AG tolerance loci for further investigation. In this study, a high-density genetic map consisting of 2711 bn markers obtained via the sequencing-based genotyping of 192 recombinant inbred lines (RILs) derived from a cross between the japonica cultivar 02428 and indica cultivar YZX was used for QTL mapping. The aims of the study were to identify stable QTLs for AG tolerance and screen candidate genes in the early and late cropping seasons (ES and LS). These QTLs will provide a foundation for further

Table 6.1: Sub-families of the rice αAmy genes.

Gene	Subfamily	Chromosome	Locus Name	CDS Coordinates (5'–3')	Regulatory Function
RAmy1 A	RAmy1	2	LOC_Os02g52700	32243146-32245056	High temperature in developing seeds
RAmy1 B		2	LOC_Os02g52710	32250180-32248279	Chemical inhibition
RAmy1 C		1	LOC_Os01g25510	14459951-14461849	High temperature in developing seeds
RAmy2 A	RAmy2	6	LOC_Os06g49970	30262778-30266915	Unknown
RAmy3 A	RAmy3	9	LOC_Os09g28400	17288993-17291295	High temperature in developing seeds
RAmy3 B		9	LOC_Os09g28420	17296166-17305076	Chemical inhibition
RAmy3 C		8	LOC_Os08g36900	23335165-23337151	Unknown
RAmy3 D		8	LOC_Os08g36910	23340676-23343533	High temperature in developing seeds Sugar starvation Calcium signaling
RAmy3 E		4	LOC_Os04g33040	20006128-2000927	High temperature in developing seedsChemical inhibition
RAmy3 F		1	LOC_Os01g51754	29760719-29770037	Unknown

investigation of the molecular mechanisms underlying anoxia tolerance and will provide target loci for improving the varieties via molecular breeding. It was also performed transcriptome expression profiling and identified differentially expressed genes located in the AG tolerance-related QTL regions, providing valuable information for candidate gene verification and the dissection of gene regulatory networks affecting rice AG tolerance. The result was phenotypic performance of the parents and RIL population for AG tolerance. The faster the rice coleoptile elongates, the sooner the seedling is able to escape the anoxic environment, which improves the chances of survival. Therefore, rice coleoptiles are a classical organ used for studies on AG tolerance. In this study, researchers investigated the dynamic phenotypic changes in the coleoptiles in two parents under anaerobic conditions at ES. Considerable distinct variations in the coleoptile traits between YZX and 02428 were observed (Fig. 6.2).

Direct seeding of rice often results in poor crop establishment due to unlevelled fields, unpredicted heavy rains after sowing, and weed and pest invasion. Thus, it is important to develop varieties which are able to tolerate flooding during germination, also known as anaerobic germination (AG), to address these constraints. A study was conducted to identify QTLs associated with AG tolerance from an IR64/Kharsu 80A F2:3 mapping population, using 190 lines phenotyped for seedling survival under the stress. Genotyping was performed using a genomewide 384-plex Indica/Indica SNP set. Four QTLs derived from Kharsu 80A providing increased tolerance to anaerobic germination were identified: three on chromosome 7 (qAG7.1, qAG7.2 and qAG7.3) and one on chromosome 3 (qAG3), with LOD values ranging from 5.7 to 7.7, and phenotypic variance explained (R2) from 8.1% to 12.6%. The QTLs identified in this study can be further investigated to better understand the genetic bases of AG tolerance in rice and used for marker-assisted selection to develop more robust direct-seeded rice varieties.

Using the software QTL Cartographer and QGene, four QTLs above the permutation threshold were identified, one positioned on chromosome 3 (qAG3.1), and the other three located on chromosome 7 (qAG7.1, qAG7.2 and qAG7.3). qAG7.2 was not detected by the CIM method, probably due to a close proximity with qAG7.3. Further, the AG1 (qAG-9-2) QTL or OsTPP7 gene that was previously reported on chromosome 9 (Angaji et al., 2010; Kretzschmar et al., 2015) was not detected in this population. In this case, AG1 could be missing in the genome of Kharsu 80A or this variety may have the susceptible allele of AG1. Several QTL mapping studies on AG tolerance in rice were published (Angaji et al., 2010; Baltazar et al., 2014; Jiang et al., 2004, 2006; Septiningsih et al., 2013b). Additionally, genome-wide association studies (GWAS) on AG tolerance have been reported (Hsu and Tung, 2015; Zhang et al., 2017). A comparative QTL study was performed to determine if QTLs in similar locations have been previously

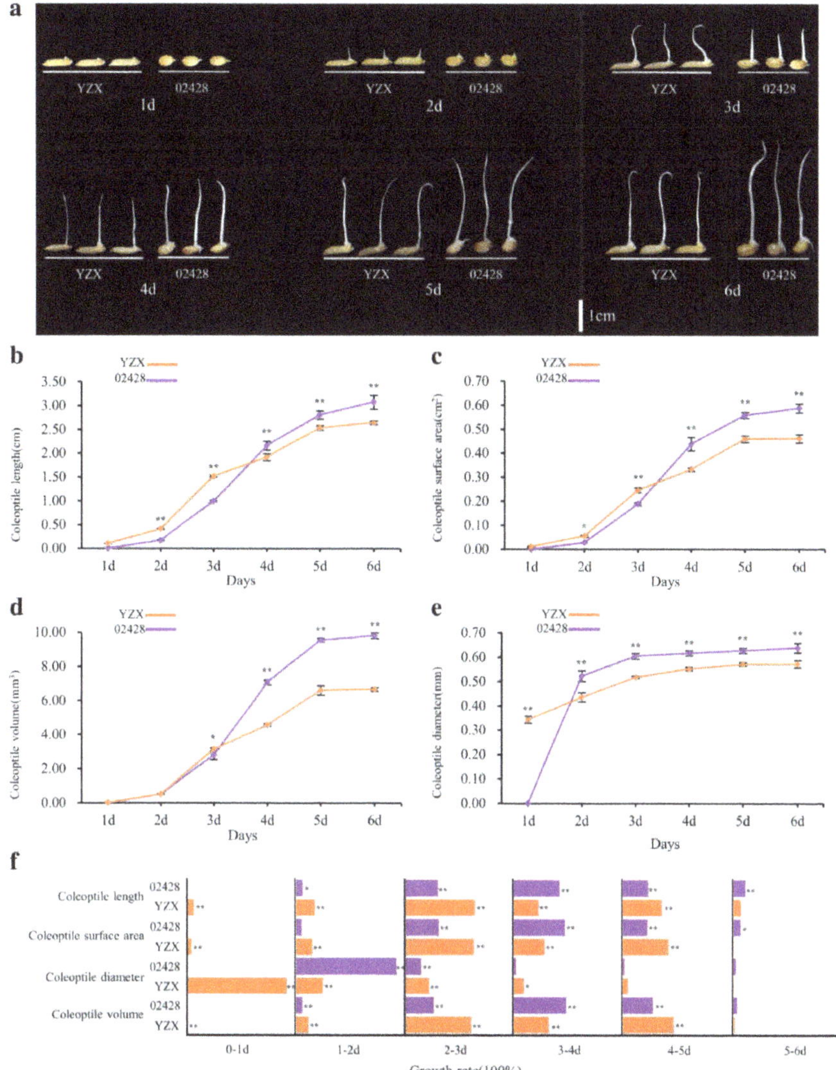

Fig. 6.2: Dynamic phenotypic changes in the coleoptiles of YZX and 02428 under anaerobic conditions. **a.** Phenotype of two parents for 1–6 d under anaerobic conditions; **b**, **c**, **d**, and **e** represent the changes in the length, surface area, diameter, and volume of the coleoptiles in 1–6 d under anaerobic conditions; **f.** The percentage daily increase in the length, surface area, diameter, and volume of the coleoptiles of YZX and 02428 under anaerobic conditions.

identified from other genetic sources. One of the QTLs on chromosome 7 (qAG7.1) derived from Kharsu 80A is in a similar position with the qAG7 identified in the IR64/Nanhi population (Baltazar et al., 2014), and partially overlapped with the largest QTL (qAG7.1) derived from Ma-Zhan (Red)

Table 6.2: QTLs for tolerance to anaerobic germination detected in an F2 population derived from IR64/Kharsu 80A population

QTL	Chr.	Flanking markers	Peak marker	QGene IM				QGene CIM				QTL Cart. IM				QTL Cart. CIM			
				LOD	R^2	Add[a]	Dom[b]	LOD	R^2	Add	Dom	LOD	R^2	Add	Dom	LOD	R^2	Add	Dom
qAG3	3	id3002377-id3004190	id3003215					5.9	18.4	8.1	4.9								
qAG7.1	7	id7000519-id7002260	wd7000465	5.5	17.5	10.7	–2.4	4.8	15.4	9.1	–0.6	5.7	12.5	9.3	6.9				
qAG7.2	7	id7002427-id7003359	d7003072	7.7	23.3	12.1	–7.5					7.1	13.5	8.7	8.5				
qAG7.3	7	id7003853-id7004429	id7004429	6.5	20.3	9.4	–4.1	5.5	17.2	7.1	–2.9	7.5	12.6	8.0	7.7	5.0	4.2	5.0	7.1

[a] Additive effects are shown for each qtL.
[b] Dominant effect are shown for each QTL.
Bold LOD scores were above the p = 0.01 threshold, normal font is above p = 0.05 threshold.

(Septiningsih et al., 2013b). Further, qAG7.2 is mapped in similar position as qAG7.2 from Ma-Zhan (Red); while qAG7.3 is in similar position with qAG7.3 derived from Ma-Zhan (Red) (Septiningsih et al., 2013b), qAG-7-2 from Khao Hlan On (Angaji et al., 2010), and a cluster of significant SNPs from two GWAS studies (Hsu and Tung, 2015; Zhang et al., 2017). The QTL on chromosome 3, qAG3, overlaps with a significant SNP from a GWAS study by Zhang et al. (2017).

It is reported that a few anaerobic germination-tolerance QTLs were found on chromosomes 5 and 11 (Jiang et al., 2004). Five different QTLs were found on chromosomes 1, 3, 7 and 9 identified by crossing IR64 with the tolerant donor Khao All Hlan On (Jiang et al., 2004; Angaji et al., 2010). The QTL qAG-9-2 or AG1 linked with local modulation of trehalose-6-phosphate (T6P) is the most promising locus for breeding (Kretzschmar et al., 2015). T6P (OsTPP7) gene is involved in metabolism of T6P (Loreti et al., 2018). In a population obtained by crossing sensitive line IR42 with tolerant variety Ma-Zhan Red (Septiningsih et al., 2013), another anaerobic germination tolerance QTL (qAG-7-1 or AG2) is found on shoot arm of chromosome 7. Zhang et al. (2017) revealed that LOC_Os06g03520 is linked with anaerobic germination tolerance using 5291 single nucleotide polymorphism markers.

Rice seeds germinating in flooded soils encounter hypoxia or even anoxia, leading to poor seed germination and crop establishment. Introgression of AG1 and AG2 QTLs associated with tolerance of flooding during germination, together with seed pre-treatment via hydro-priming or pre-soaking can enhance germination and seedling growth in anaerobic soils. In a study, researchers assessed the performance of elite lines incorporating AG1, AG2 and their combination when directly seeded in flooded soils using dry seeds. The QTLs were in the background of two popular varieties, PSB Rc82 and Ciherang-Sub1, evaluated along with the donors Kho Hlan On (AG1) and Ma-Zhan Red (AG2) and recipient parents PSB Rc82 and Ciherang-Sub1. In one set of experiments conducted in the greenhouse, seedling emergence, growth, and carbohydrate mobilization from seeds were assessed. Metabolites associated with reactive oxygen species (ROS) scavenging, including malondialdehyde (MDA) as a measure of lipid peroxidation, ascorbate, total phenolic concentration (TPC), and activities of ROS scavenging enzymes were quantified in seeds germinating under control (saturated) and flooded (10cm) soils. In another set of experiments conducted in a natural field with 3–5cm flooding depths, control and pre-treated seeds of Ciherang-Sub1 introgression lines and checks were used. Flooding reduced seedling emergence of all genotypes, though emergence of AG1+AG2 introgression lines was greater than the other AG lines. Soluble sugars increased, while starch concentration decreased gradually under flooding, especially in the tolerant checks and in AG1+AG2 introgression lines. Less lipid peroxidation and higher

Table 6.3: List of QTLs detected for anaerobic germination potential in the BC1F2:3 population of Kalarata and NSIC Rc238.

QTL NAME	CHR	PEAK MARKER	POS (cM)	INT (cM)	LOD	PVE	ADD
Survivability at 14 DAS							
Screenhouse conditions							
$qSUR3\text{-}1_{Rc238\text{-}SCR\text{-}14}$	3	SWRm_00276	190.0	178–198	4.90	11.06	13.04
$qSUR5\text{-}1_{Rc238\text{-}SCR\text{-}14}$	5	SWRm_00427	108.0	104–118	4.50	10.82	–12.47
$qSUR6\text{-}1_{Rc238\text{-}SCR\text{-}14}$	6	SWRm_00495	36.0	32–46	6.00	14.38	11.66
$qSUR7\text{-}1_{Rc238\text{-}SCR\text{-}14}$	7	SWRm_01153	62.0	59–66	18.40	36.68	20.13
$qSUR8\text{-}1_{Rc238\text{-}SCR\text{-}14}$	8	SWRm_00599	66.2	64–74	5.20	12.04	–14.38
Tray-on-table conditions							
$qSUR3\text{-}1_{Rc238\text{-}TAB\text{-}14}$	3	SWRm_00276	192.0	182–198	5.80	12.79	10.85
$qSUR5\text{-}1_{Rc238\text{-}TAB\text{-}14}$	5	SWRm_00427	108.0	104–116	4.10	9.76	–13.09
$qSUR6\text{-}1_{Rc238\text{-}SCR\text{-}14}$	6	SWRm_00495	46.0	38–50	6.60	15.15	9.79
$qSUR7\text{-}1_{Rc238\text{-}TAB\text{-}14}$	7	SWRm_01153	60.0	58–66	15.30	31.58	17.15
$qSUR8\text{-}1_{Rc238\text{-}TAB\text{-}14}$	8	SWRm_00599	66.2	64–74	4.90	11.70	–13.14
Survivability at 21 DAS							
Screenhouse conditions							
$qSUR3\text{-}1_{Rc238\text{-}SCR\text{-}21}$	3	SWRm_00276	194.0	182–202	4.87	11.04	14.03
$qSUR5\text{-}1_{Rc238\text{-}SCR\text{-}21}$	5	SWRm_00427	110.0	106–116	4.49	10.77	–14.8
$qSUR6\text{-}1_{Rc238\text{-}SCR\text{-}21}$	6	SWRm_00495	38.0	34–44	5.97	14.49	12.72
$qSUR7\text{-}1_{Rc238\text{-}SCR\text{-}21}$	7	SWRm_01153	62.0	59–66	20.31	39.66	24.28
$qSUR8\text{-}1_{Rc238\text{-}SCR\text{-}21}$	8	SWRm_00599	66.2	64–76	5.21	12.03	–16.00
Tray-on-table conditions							
$qSUR3\text{-}1_{Rc238\text{-}TAB\text{-}21}$	3	SWRm_00276	192.0	182–198	5.55	12.32	11.79
$qSUR5\text{-}1_{Rc238\text{-}TAB\text{-}21}$	5	SWRm_00427	110.0	106–116	4.57	10.81	–16.90
$qSUR6\text{-}1_{Rc238\text{-}TAB\text{-}21}$	6	SWRm_00495	44.0	38–48	6.96	16.01	12.11
$qSUR7\text{-}1_{Rc238\text{-}TAB\text{-}21}$	7	SWRm_01153	62.0	58–66	17.50	35.17	22.76
$qSUR8\text{-}1_{Rc238\text{-}TAB\text{-}21}$	8	SWRm_00599	66.2	64–74	4.78	11.28	–15.5

q indicates QTL, *SCR* screenhouse, *TAB* tray-on-table, *POS* position, *INT* position interval, *LOD* logarithm of odds, *PVE* percent phenotypic variation explained, *ADD* additive effects of the peak marker.

α-amylase activity, higher ascorbate (RAsA) and TPC were observed in the tolerant checks and in the AG1+AG2 introgression lines. Lipid peroxidation correlated negatively with ascorbate, TPC, and with ROS scavengers. Seed hydro-priming or pre-soaking increased emergence by 7–10% over that of dry seeds. Introgression of AG2 and AG1+AG2 QTLs with seed pre-treatment showed 101–153% higher emergence over dry seeds of intolerant genotypes in the field. Lines carrying AG1+AG2 QTLs showed higher α-amylase activity, leading to rapid starch degradation and increase in soluble sugars, ascorbate, and TPC, together leading to higher germination and seedling growth in flooded soils. Seed hydro-priming or pre-soaking for 24 hours also improved traits associated with flooding tolerance. Combining tolerance with seed management could therefore, improve crop establishment in flooded soils and encourage large-scale adoption of direct seeded rice system. In a study, it was observed that the de-husked grain of Ciherang-Sub1-AG1-AG2, Ciherang-Sub1-AG2, PSB Rc82-AG1-AG2, and the tolerant checks KHO and MR is brownish in colour, which is coupled with higher emergence and faster growth, as well as with other physiological traits positively associated with tolerance.

Anaerobic germination (AG) is an important trait for direct-seeded rice (DSR) to be successful. Rice usually has low germination under anaerobic conditions, which leads to a poor crop stand in DSR when rain occurs after seeding. The ability of rice to germinate under water reduces the risk of poor crop stand. Further, this allows the use of water as a method of weed control. The identification of genetic factors leading to high anaerobic germination is required to develop improved DSR varieties. In the present study, two BC1F2:3 mapping families involving a common parent with anaerobic germination potential, Kalarata, an indica landrace, and two recurrent parents, NSIC Rc222 and NSIC Rc238, were used. Phenotyping was done under two environmental conditions and genotyping was carried out through the KASP SNP genotyping platform. A total of 185 and 189 individuals, genotyped with 170 and 179 polymorphic SNPs, were used for QTL analysis of the two populations, Kalarata/NSIC Rc238 and Kalarata/NSIC Rc222, respectively. A total of five QTLs on chromosomes 3, 5, 6, 7, and 8 for survival (SUR) and four QTLs on chromosomes 1, 3 (two locations), and seven for the trait seedling height (SH) across the populations and over the screening conditions were identified. Except for the QTLs on chromosomes 5 and 8, the parent with AG potential, Kalarata, contributed all the other QTLs. Among the five QTLs for SUR, the second-largest QTL (qSUR6–1) was novel for AG potential in rice, showing a stable expression in terms of genetic background and screening conditions explaining 11.96% to 16.01% of the phenotypic variation. The QTL for SH (qSH1–1) was also novel. Considering different genetic backgrounds and different screening conditions, the QTLs identified for the trait SUR explained phenotypic variation in the range of 57.60% to 73.09% while that for the trait SH ranged from 13.53% to 34.30%.

6.2 QTLs for Submergence Tolerance

The major landmark in the history of submergence-tolerant breeding was the identification of QTL controlling the submergence, which was designated as the Sub1 gene (Xu and Mackill, 1996). Identification of the genomic segment encoding the Sub1 QTL had been narrowed down to a small chromosomal region. The gene was later cloned and identified to be an ethylene response factor (ERF) (Xu et al., 2006). This genomic region was further mapped to a 0.16 cM region on chromosome 9, with approximately 3000 F2 progeny (Xu et al., 2000). Subsequently, using a fine-scale physical mapping study, Xu et al. (2006) narrowed the locus down to 0.075 cM using 4022 F2 individuals. Septiningsih et al. (2012) identified four novel QTLs on chromosomes 1, 2, 9, and 12 in a cross between IR72 and Madabaru (both moderate submergence-tolerant varieties) at F2:3 populations. A subset of 80 families was selected from the two extreme sides for the QTL analysis from the F2:3 population of 466 families using phenotypic screening under submergence stress. Also, three non-Sub1 QTLs from IR72 were identified, suggesting that alternative pathways may exist independent of the ethylene-dependent pathway of the Sub1 gene (Septiningsih et al., 2012).

Sripongpangkul et al. (2000) performed QTL analysis in a recombinant inbred lines (RILs) cross of IR74/Jalmagna using leaf length, internode length, and plant height increment to determine the stem elongation mechanism in deep-water rice, and detected 26 QTLs regulating submergence tolerance and plant elongation. Using the rate of internode elongation (RIE) and lowest elongated internode (LEI), Kawano et al. (2008) detected two QTLs on chromosomes 1 and 12, and 3 and 12 for each trait, respectively. Similarly, Nemoto et al. (2004) detected two QTLs on chromosomes 3 and 12, using the LEI trait. Using three parameters, REI, LEI, and total internode elongation length (TIL) on three different populations, Hattori et al. (2007) identified three QTLs on 1, 3, and 12 in similar genomic regions in different populations. These, however, suggest that such QTLs are constantly responsible for elongation in response to flooding in deep-water rice. Furthermore, Hattori et al. (2008) reported QTL on chromosome 12, which is most effective for internode elongation in deep-water using near-isogenic lines (NILs) on genetic background of Taichung65 (NIL-1, NIL-3, and NIL-12). In deep-water rice, elongation under water is considered the most important phenotyping trait in QTL analysis. Using this trait, Nagai et al. (2012) identified two new QTLs, qTIL2 and qTIL4, controlling total internode elongation during the seedling stage. Research studies on the identification of QTLs controlling different types of submergence tolerance were summarized in Table 6.4. The mapping of major QTLs associated with anaerobic germination (AG) has been carried out and identified at the IRRI (Angaji et al., 2010; Hattori et al., 2008; Nagai et al., 2012; Baltazar et al., 2015) among these QTLs, qAG-9-2 was mapped to circa 50 kb on chromosome 9 using the Khao Hlan On rice variety (Angaji et al., 2010). The gene encoding a trehalose-6-phosphate phosphatase (OsTPP) conferring stress tolerance has been

Table 6.4: Summary of quantitative trait locus for submergence tolerance traits in rice.

Cross	Trait	Population	Chromosome	References
Nanhi × IR64	anaerobic germination	F_2, F_3	2, 7, 11	Septiningsih et al., 2012
Bhaduaf × Taichung 65	lowest elongated internode, rate of internode elongation	F_2	1, 3,12	Hamamura and Kupkanchankul, 1979
FR13A × KDML105	vegetative stage submergence	Backcross line	1, 7, 9	Siangliw et al., 2006
IR42 × FR13A	Survival rate	recombinant inbred line	1, 4, 8, 9, 10	Gonzaga et al., 2016
USSR5 (japonica) x N22 (indica)	Germination under low temperature and anoxia	F_2	5,11	Jiang et al., 2006
1R40931-26 × PI543851	vegetative stage submergence	F_2	9	Xu and Mackill, 1996
Jalmagna × IR74	vegetative stage submergence	recombinant inbred line	9	Ramaiah and Ramaswami, 1940
Habiganj Aman × Patnai23	lowest elongated internode	F_2	1, 12	Tang et al., 2005
1R49830-7-1-2 × CT6241-17-1-5-1	vegetative stage submergence	Doubled Haploid Line, recombinant inbred line, F_2	9	Toojinda et al., 2003
T 65 × Bhadua	total internode length	recombinant inbred line	2, 4	Xu et al., 2000
TX9425 x Naso Nijo	Membrane potential and hypoxia	Doubled Haploid Line	2	Gill et al., 2017
Madabaru × IR72	vegetative stage submergence	F_2, F_3	1, 2, 9, 12	Septiningsih et al., 2012
Goai × Patnai 23	lowest elongated internode	F_2	3,12	Tripathi and Rao, 1985
Ma-Zhan (Red) × IR42	anaerobic germination	Backcross line	2, 5, 6, 7	Septiningsih et al., 2013
FR13A × IR74	vegetative stage submergence	recombinant inbred line	6, 7, 9, 11, 12	Nandi et al., 1997
Khao Hlan On × IR64	anaerobic germination	Backcross line	1, 3,7, 9	Angaji et al., 2010

Table 6.4 cont. ...

...Table 6.4 cont.

Cross	Trait	Population	Chromosome	References
Nipponbare × IR64	coleoptile elongation	recombinant inbred line	1	Hsu and Tung, 2015
Nipponbare/ Kasalath// Nipponbare	Submergence tolerance	Backcross line	1, 3, 4, 6, 7	Manangil et al., 2013
DX18-121 X M-202	Submergence tolerance	F_2	9	Xu et al., 2000
C9285 × T65	lowest elongated internode, total internode length, number of elongated internode	Backcross line	1,3,12	Suge, 1987

cloned and functionally validated (Kretzschmar et al., 2015). This QTL has also been transferred to popular high-yielding rice varieties (Toledu et al., 2015).

In order to establish a more effective marker-assisted selection (MAS) breeding scheme for introgression of flood-tolerance traits from landraces to commercial varieties, it is essential to find additional DNA diagnostic markers that are tightly linked with the traits of interest, which will primarily come from a fine mapping of the identified QTLs. Isolation of these QTLs underlying genes will help in designing more accurate gene-specific DNA markers for MAS and would also advance our understanding of the physiological and genetic mechanisms of flood tolerance. Furthermore, such research would facilitate stronger and new alleles for tolerance. However, the application of MAS in quantitative traits improvement in plant breeding is hindered by the low predictive value of QTL markers for performance. This is caused by numerous factors, such as strong genotype by environment interaction (Oladosu et al., 2017), low expression of certain genes, and recombination between markers and target genes (Paterson et al., 1991; Young, 1996). Genome sequencing projects have made it possible to use known candidate gene sequences as markers to associate with target traits. The candidate gene approach has emerged as a promising method of merging QTL analysis with the extensive data available on the cloning and characterization of genes involved in plant stressors (Faris et al., 1999).

Flash flooding of young rice plants is a common problem for rice farmers in South and Southeast Asia. It severely reduces grain yield and increases the unpredictability of cropping. The inheritance and expression of traits associated with submergence stress tolerance at the seedling stage are physiologically and genetically complex. It was exploited naturally occurring differences between certain rice lines in their tolerance to submergence and used quantitative trait loci (QTL) mapping to improve understanding of the genetic and physiological basis of submergence tolerance. Three rice populations, each derived from a single cross between two cultivars differing in their response to submergence, were used to identify QTL associated with plant survival and various linked traits. These included total shoot elongation under water, the extent of stimulation of shoot elongation caused by submergence, a visual submergence tolerance score, and leaf senescence under different field conditions, locations and years. Several major QTL determining plant survival, plant height, stimulation of shoot elongation, visual tolerance score and leaf senescence, each mapped to the same locus on chromosome 9. These QTL were detected consistently in experiments across all the years and in the genetic backgrounds of all three mapping populations. Secondary QTL influencing tolerance were also identified and located on chromosomes 1, 2, 5, 7, 10 and 11. These QTL were specific to particular traits, environments, or genetic backgrounds. All identified QTL contributed to increased submergence tolerance through their effects on decreased underwater shoot elongation or increased maintenance of chlorophyll levels, or on both. These findings establish the foundations of a marker-assisted scheme for introducing submergence tolerance into agriculturally desirable cultivars of rice.

QTL for traits responsive to submergence in all three mapping populations mapped to rice chromosomes 1, 2, 5, 7, 9, 10 and 11 (Table 6.4). In general, the analysis indicated that genetic control of submergence tolerance in rice is more complex than reported earlier by Xu and Mackill (1996). Major QTL controlling traits associated with PPS, TSE, RSE, TS and LS, all mapped to the same region of chromosome 9 in each of the three mapping populations and were consistently identified in most of, if not all, the experiments. This indicates that one QTL on chromosome 9 (QTLch9) is probably the major genetic contributor to submergence-tolerance performance in these three rice populations. The QTLch9 was detected with both SIM and sCIM procedures of MQTL. Significant QTL that exceeded the significance threshold are shown in Fig. 6.4. They mapped to the RM41±RG553 interval on the short arm of chromosome 9. This QTL accounted for 48, 63, 74 and 72 % of phenotypic variation explanation (PVE) for PPS, TS, TSE and LS, respectively, in the DHL (Table 6.4).

The expression of submergence tolerance is known to be environmentally dependent and genetically complex (Suprihatno and Coffman, 1981; Mohanty and Khush, 1985; Sinha and Saran, 1988; Haque et al., 1989; Adkin et al., 1990; Setter et al., 1997). In different years and seasons and with different mapping populations, the QTL controlling traits related to submergence tolerance were mapped on many genomic regions. However, the consistently detected QTLch9 indicated a common genetic factor controlling submergence tolerance in rice. Therefore, QTLch9 is the most important submergence tolerance QTL in the three populations. The QTLch9 was mapped near the Sub1 locus previously identified by Xu and Mackill (1996) and Nandi et al. (1997). The coincidence of the map locations of all traits on chromosome 9 may be due either to linkage or to pleiotropy. At the level of mapping resolution afforded by three populations, linkage and pleiotropy cannot be distinguished.

The coincidence of QTL for traits important for submergence tolerance provides significant genetic evidence linking physiological responses to submergence tolerance in rice. Multiple-season QTL analysis in three mapping populations has enabled us to identify a major QTL and several minor QTL that are of potential value in the improvement of submergence tolerance in rice. Availability of cloned genes present at these loci will be important stepping stones towards a fuller understanding of the regulation of submergence tolerance. Map-based cloning of the major QTL for submergence tolerance on chromosome 9 holds the key to improving our understanding of the physiological basis of submergence tolerance in rice and developing dependable marker-aided selection schemes for breeding submergence-tolerant cultivars with desirable agronomic features.

6.3 Sub1 QTL Analysis

The quantitative trait locus (QTL) primarily responsible for the tolerance of FR13A was mapped to a small region called Sub1 (Xu and Mackill, 1996; Xu et al., 2006; Hattori et al., 2009). Flood-intolerant rice varieties become tolerant after introgression of the Sub1 locus, demonstrating that the Sub1 QTL is sufficient for flood tolerance (Xu and Mackill, 1996). The Sub1 locus from FR13A and bioengineered tolerant genotypes with the introgressed Sub1 locus has three Ethylene Response Factor (ERF) genes: Sub1A, Sub1B and Sub1C (Xu et al., 2006). All three genes belong to a large transcription factor family referred to as ERFs and they group within the abiotic stress responding ERF subfamily VII (Nakano et al., 2006; Jung et al., 2010). The intolerant parent variety M202 used in previous submergence studies and in the experiments discussed here has the Sub1B and Sub1C genes but lacks the Sub1A gene. The tolerant near isogenic line used in experiments is the M202-Sub1 that has the Sub1 locus from FR13A containing the Sub1A gene that confers submergence tolerance (Fukao et al., 2006; Xu et al., 2006). Upon complete submergence, Sub1A is rapidly induced within a day. Expression of Sub1C is induced more strongly in the intolerant M202 when compared to the tolerant M202-Sub1. Sub1B transcript abundance was reported to be induced by submergence in both genotypes but its transcript abundance was slightly higher in M202-

Sub1 compared to M202 (Fukao et al., 2006). Expression of the ubiquitin promoter driven Sub1A transgene (LG (Sub1A)) in an intolerant variety, Liaogeng (LG) was sufficient.

Until the mid-1990s, the genetic control of submergence tolerance remained ambiguous. Several studies suggested that it was a typical quantitative trait (Suprihatno and Coffman, 1981; Mohanty et al., 1981; Mohanty and Khush, 1985; Haque et al., 1989). Molecular mapping allowed the identification of the major QTL SUBMERGENCE 1 (Sub1) on chromosome 9, contributing up to 70% of phenotypic variation in tolerance (Xu and Mackill, 1996). Several independent studies confirmed the major chromosome 9 QTL and identified other minor QTLs that accounted for less than 30% of the phenotypic variation in tolerance (Nandi et al., 1997; Toojinda et al., 2003; Siangliw et al., 2003; Fig. 6.3). Sub1 was further mapped with circa 3,000 F2 progeny to a 0.16-cM region on chromosome 9 (Xu et al., 2000). This was followed by a fine-scale physical mapping study using 4,022 F2 individuals, narrowing the locus to 0.075 cM (150 kb; six recombinants; Xu et al., 2006). The positional cloning of Sub1 was accomplished by identifying a contiguous set of BAC and binary clones spanning the region, using the intolerant indica Teqing and the tolerant FR13A derivative IR40931-26, respectively. This resolved a chromosomal integral from marker CR25K to SSR1A that varies in length, structure, and gene content in the japonica and indica genomes, using Nipponbare and Teqing as representatives. The sequencing of the Teqing BAC contig in the Sub1 region confirmed the presence of ~ 50% interspersed sequences of transposon or retrotransposon origin (Xu et al., 2006). Recombination suppression in this region was associated with an inversion and large deletion between markers 101O9L and 14A11-71K, in japonica relative to submergence intolerant Teqing and tolerant IR40931-26. This variant segment of Sub1 contains a gene-designated Sub1A that encodes a putative ERF DNA binding protein with a single ERF/APETELA2 domain (Xu et al., 2006; Fig. 6.3). Two additional ERFs, designated Sub1B and SUB1C, were identified proximal to the centromere within ~ 100 bp of Sub1A in Teqing. The genome of the submergence intolerant japonica cultivar Nipponbare encodes Sub1B and Sub1C in this chromosomal interval but lacks Sub1A (GenBank AP006758 and AP005705). A survey of the ERF gene allele composition at the Sub1 locus, or Sub1 haplotype, of 21 rice accessions confirmed that all encode Sub1B and Sub1C, but only a sub-set of indica and aus accessions encode Sub1A (Xu et al., 2006). Phylogenetic analysis of the Sub1 ERFs indicates that Sub1A arose from a duplication of Sub1B, possibly after the domestication of indica rice (Fukao et al., 2009). Sub1A is recognized in two allele forms in submergence tolerant and intolerant indica and aus accessions, based on nucleotide variations in the protein-coding region (Xu et al., 2006). SubA1-1 is found only in tolerant lines, such as FR13A, whereas the Sub1A-2 allele is present in

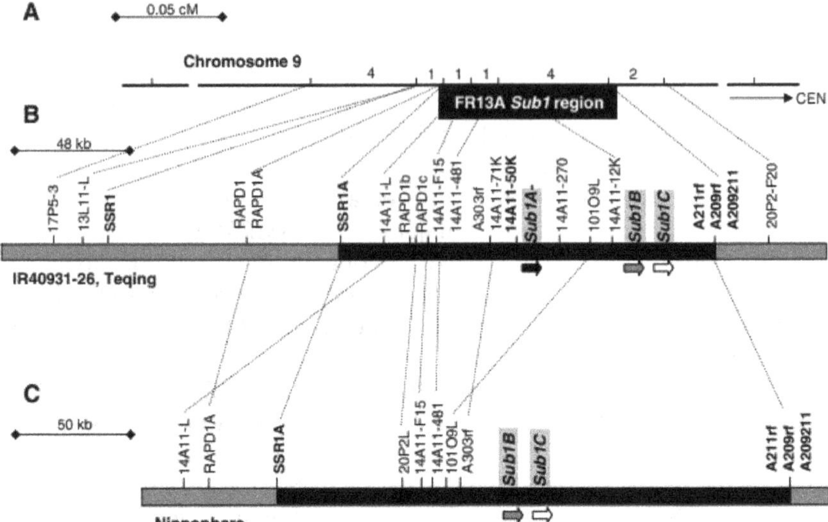

Fig. 6.3: Genetic and physical map of the Sub1 locus on the long arm of chromosome 9. High-resolution maps represent FR13A derivative IR40931-26, Teqing (indica) and Nipponbare (japonica). The vertical lines indicate markers between maps. (a) The chromosomal segment defined by recombinant mapping as the Sub1 region, and its orientation relative to the centromere is indicated. Amplified fragment length polymorphism markers and their derivatives are prefixed with the letter 'A'. Genetic distances between adjacent markers are measured using the number of recombinant plants (one recombinant plant = 0.0012 cM). (b) Physical map spanning the Sub1 locus constructed by establishment and analysis of a BAC contig of six clones, including one from IRBB21 and five from Teqing, both intolerant indica. A contig of binary clones from IR40931-26, a submergence tolerant FR13A derivative, was also assembled across this region. The Sub1 locus was delimited to 42 kb between SSR1A and 14A11-481 and 112 kb between 14A11-f15 and AFLP209rf. Three group VII ERF genes are encoded in this region in IR04931-26 and Teqing, Sub1A, Sub1B, and Sub1C. IR40931-26 has the Sub1A-1 allele and Teqing the Sub1A-2 allele. (c) Physical map of the Sub1 region in the Nipponbare genome. Based on DNA sequence data from GenBank, colinearity with the two physical maps (b, c) holds for most markers, Sub1B and Sub1C. However, there is an inversion and deletion of the region with the landmark marker 101O9L and Sub1A. The sequenced Sub1 region in Nipponbare is 142 kb (Adapted from Supplementary Fig. 1 of Xu et al., 2006).

intolerant indica accessions. Sub1A-1 and Sub1A-2 encode identical proteins, with the exception of Ser186 in the tolerant allele and Pro186 in the intolerant allele. Another distinction is recognized at the level of gene expression. Sub1A-1 promotes rapid, prolonged, and pronounced transcript accumulation in leaves of 14- to 28-day-old plants in response to submergence, whereas Sub1A-2 promotes a lower level of transcript induction by the stress (Fukao et al., 2006; Xu et al., 2006). Importantly, strong ectopic expression of the Sub1A-1 cDNA from FR13A in a highly intolerant japonica proved to be sufficient to confer robust submergence tolerance (Xu et al., 2006). An extensive analysis of diverse rice accessions with different Sub1 haplotype showed that submergence tolerance is associated with variable levels of Sub1A transcript in internodes and nodes at the heading stage (Singh et al., 2010). In Nipponbare transgenics expressing the GUS reporter gene under the control of the native Sub1A promoter, GUS activity was specifically induced by submergence in regions of leaves associated with growth, i.e. the leaf base and leaf collar, suggesting that Sub1A might act on the processes of cell division and/or cell elongation in leaves under submergence.

Submergence-tolerant rice maintains viability during complete submergence by limiting underwater elongation until flood-waters recede. Acclimation responses to submergence are coordinated by the submergence-inducible *Sub1A*, which encodes an ethylene-responsive factor-type transcription factor (ERF). *Sub1A* is limited to tolerant genotypes and is sufficient to confer submergence tolerance to intolerant accessions.

In the *Sub1A* overexpression line, SLR1 protein levels declined under prolonged submergence. Submergence-tolerant rice maintains viability during complete submergence by limiting underwater elongation until flood-waters recede. Acclimation responses to submergence are coordinated by the submergence-inducible *Sub1A*, which encodes an ethylene-responsive factor-type transcription factor (ERF). *Sub1A* is limited to tolerant genotypes and is sufficient to confer submergence tolerance to intolerant accessions. Here we evaluated the role of *Sub1A* in the integration of ethylene, abscisic acid (ABA), and gibberellin (GA) signalling during submergence. The submergence-stimulated decrease in ABA content was *Sub1A*-independent, whereas GA-mediated underwater elongation was significantly restricted by *Sub1A*. Transgenics that ectopically express *Sub1A* displayed classical GA-insensitive phenotypes, leading to the hypothesis that *Sub1A* limits the response to GA. Notably *Sub1A* increased the accumulation of the GA signalling repressors Slender Rice-1 (SLR1) and SLR1 Like-1 (SLRL1) and concomitantly diminished GA-inducible gene expression under submerged conditions. In the *Sub1A* overexpression line, SLR1 protein levels declined under prolonged submergence but were accompanied by an increase in accumulation of SLRL1, which lacks the DELLA domain. In the presence of *Sub1A*, the increase in these GA signalling repressors and decrease in GA responsiveness were stimulated by ethylene, which promotes *Sub1A* expression. Conversely, ethylene promoted GA responsiveness and shoot elongation in submergence-intolerant lines. Together, these results demonstrate that *Sub1A* limits ethylene-promoted GA responsiveness during submergence by augmenting accumulation of the GA signalling repressors SLR1 and SLRL1.

A key feature of the Sub1A-mediated response is restricting rice shoot growth during complete submergence to conserve energy until flood-waters recede. Shoot growth is driven by cell elongation and/or cell division and responds to environmental factors, such as light, temperature, and hormones like gibberellins, auxin, and brassinosteroids. The Sub1A response was linked to GA signalling, resulting in decreased shoot elongation when submerged (Fukao and Bailey-Serres, 2008). Results suggest that brassinosteroids could be involved in Sub1A-mediated dampening of the GA responses during submergence. BR significantly restricts shoot elongation in M202 plants during submergence, an effect similar to the SUB1A-restricted shoot elongation in M202-Sub1 plants (Fig. 6.4).

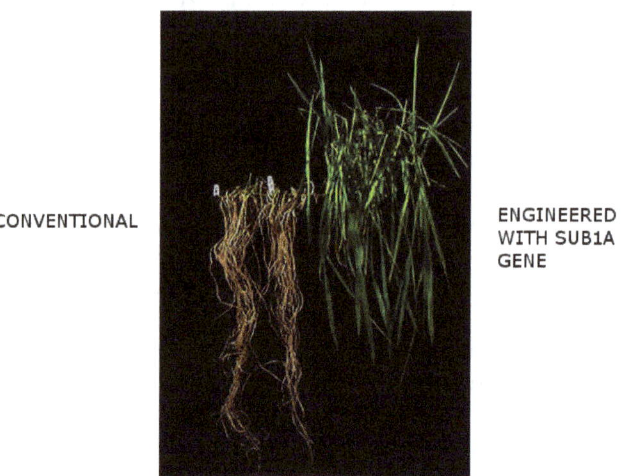

Fig. 6.4: After receding submergence the conventional and Sub1A rice.

6.4 SD1 Linkage Mapping

A combination of GWASs and linkage mapping recently identified SEMIDWARF1 (SD1) as a causal gene for the deepwater response underlying the QTL on chromosome 1 (Kuroha et al., 2018). SD1 encodes *O. sativa* gibberellin 20 oxidase 2 (OsGA20ox2), an enzyme involved in GA biosynthesis. Loss-of-function of SD1 confers a semi-dwarf phenotype with a short internode (culm) and was responsible for a higher harvest index and greater grain yield during the Green Revolution (Sasaki et al., 2002). Unlike the loss-of-function haplotype, the deepwater rice–specific SD1 haplotype (SD1- DW) contributes to rapid elongation of the internodes in submerged rice via a specific mode of regulation referred to as the 'ethylene-gibberellin relay' (Kuroha et al., 2018) (Fig. 6.5). In deepwater rice under submergence, the OsEIL1a protein stabilized by ethylene directly transactivates the SD1-DW promoter; transactivation by ethylene is not observed in non-deepwater rice. The 17 polymorphisms specific to the SD1-DW haplotype located at the promoter and second intron lead to this ethylene-inducible transactivation via an unknown mechanism (Kuroha et al., 2018). SD1-DW prefers production of the active GA species, GA_4, rather than GA_1. Since GA_4 has a stronger effect on internode elongation than GA_1, efficient and rapid GA_4 production by both high transactivation and enzymatic activity of SD1-DW contributes to rapid internode elongation of deepwater rice in response to submergence (Kuroha et al., 2018). While two exonic polymorphisms of SD1-DW related to the high enzymatic activity are conserved in almost all *O. sativa* varieties, excluding Japonica spp. (Asano et al., 2011; Zhao et

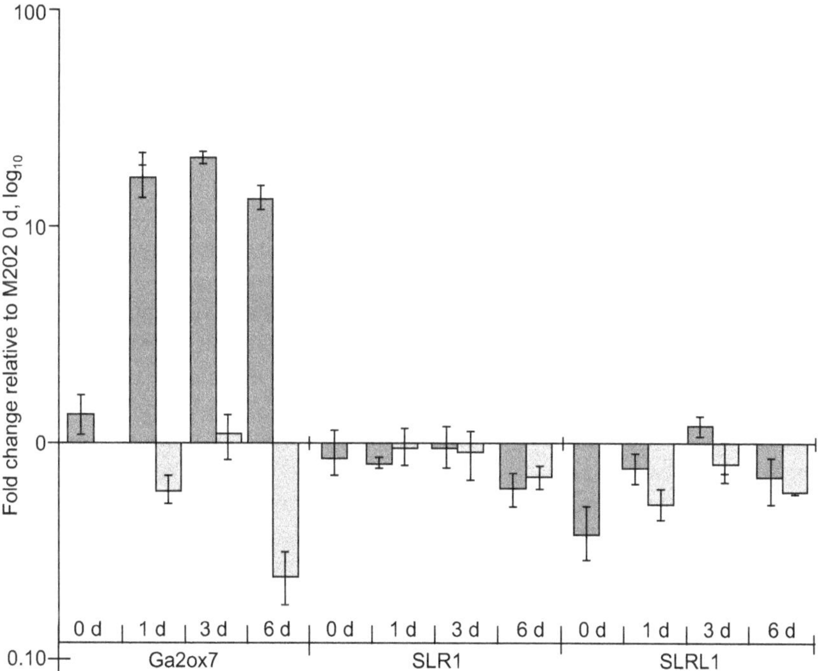

Fig. 6.5: Expression profiles of gibberellic acid (GA) catabolic and signalling genes in a time-course experiment. Expression levels for GA2ox7, SLR1 and SLRL1 at 0, 1, 3, and 6 d after submergence in M202 (light gray bars) and M202-Sub1 (dark gray bars) were measured using real-time PCR assays. Values are relative to M202 0 d. Error bars indicate SE. Notably, expression of GA2ox7 was repressed in the intolerant M202 upon submergence in contrast with induction in M202-Sub1.

al., 2018), the 17 polymorphisms for high transactivation are specific to only some deepwater rice varieties in Bangladesh (Kuroha et al., 2018). Varieties harbouring both SD1-DW and SK1/2 display stronger deepwater responses than others, suggesting that the SD1-DW-mediated 'ethylene-gibberellin relay' system works synergistically with the presence of SK1/2 to enhance the deepwater response under submergence (Fig. 6.6).

Unlike deepwater rice, non-deepwater rice cannot perform internode elongation at the vegetative stage, even with GA treatment (Nagai et al., 2014; Kuroha et al., 2018). Since at least three QTLs distinct from SD1-DW and SK1/2 were detected for GA responsiveness (Nagai et al., 2014), it is likely that unidentified factors are involved in GA responsiveness for internode elongation in deepwater rice. It has recently been reported that the transcription factor Premature Internode Elongation 1 (PINE1) regulates internode elongation by increasing GA responsiveness (Gomez-Ariza et al., 2019). Identification of QTLs for GA responsiveness in deepwater rice and their relationships with PINE1 may help pave the way to new approaches to understanding the GA-mediated deepwater response.

Although complete submergence is a common natural disaster that damages rice production in many rice growing areas throughout the world, all commercially important cultivars are intolerant to the stress. The identification of the Sub1 QTL enabled its transfer by marker-assisted backcrossing (MABC) into the farmer preferred varieties (Xu et al., 2004; Mackill, 2006). The gene-level analyses of the Sub1 region resolved single nucleotide polymorphisms within Sub1A and Sub1C that could be used for molecular markers and in precision breeding (Neeraja et al., 2007; Septiningsih et al., 2009). Using MABC, a small genomic region containing Sub1A has been introgressed into modern high-yielding varieties, such as Swarna, Samba Mahsuri, IR64, Thadokkam 1 (TDK1), CR1009, and BR11 (Septiningsih et al., 2009). Microsatellite markers that were polymorphic between the two parents were used to ensure that the recurrent parent genome was combined with the Sub1 region originally from FR13A on chromosome 9. Multiple evaluations of submergence tolerance in the greenhouse and farmers' fields confirmed that all 'Sub1' lines exhibit significantly greater tolerance to complete submergence as compared with their original parents (Sarkar et al., 2009; Septiningsih et al., 2009; Singh et al., 2009; Fig. 6.6).

These studies indicate that the introgression of the Sub1 region of FR13A through MABC is widely applicable to diverse genetic backgrounds. In addition to submergence tolerance, the effect of Sub1 on growth, maturation, grain production, and grain quality was assessed in Swarna-Sub1, IR64-Sub1, and Samba Mahsuri-Sub1. Comparative analysis of the three Sub1 varieties and their original parents revealed that introgression of SUB1 does not negatively affect agronomical performance, including

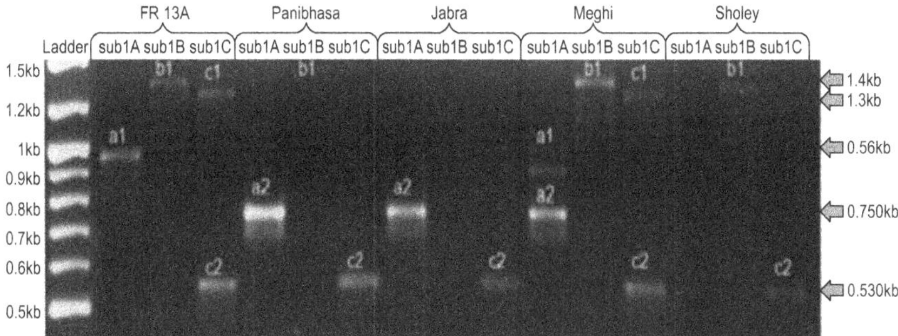

Fig. 6.6: Gel picture showing amplified bands for the Sub1 loci among the studied rice lines.

yield and grain quality, under regular growth conditions (Sarkar et al., 2006, 2009; Neeraja et al., 2007; Singh et al., 2009). In intolerant varieties, complete inundation at the vegetative stage considerably decreases the number of panicles, number of grains per panicle, and grain-filling percentage and delayed flowering and maturity, causing a dramatic decline in grain yield. Sub1 rice minimizes the reduction of these reproductive traits by a submergence event and produces three- to six-fold more grain by weight than the non-Sub1 varieties (Singh et al., 2009).

6.5 Mapping and Identification of Molecular Sub1 Locus on Chromosome 9S

Sub1 region was delineated by recombination and molecular markers to be ~ 200 kb. A single gene in this interval was identified as Sub1 gene candidate. The presence of three ethylene responsive factor-domain containing genes designated Sub1A, 1B and 1C was confirmed. The corresponding region of the publicly available Nipponbare (japonica) genome spans 225 kb and is rearranged, relative to the indica genome. The ERF gene Sub1A is absent from this region in Nipponbare. Sub1A and 1C mRNAs strongly but transiently under submergence are reduced upon de-submergence in seedling leaves. A test of Sub1 haplotype in 17 indica and four japonica varieties identified two Sub1A, five 1B and eight 1C alleles on the basis of variation in amino acid sequence. Submergence tolerance was correlated with presence of submergence-induced Sub1A-1, and intolerance was associated with Sub1A-2 or complete absence of this gene. To validate the role of Sub1A-1, an intolerant japonica variety, Liaogeng, was transformed with Sub1A-1 full-length cDNA under the control of the maize ubiquitin1 promoter. Independent transgenic lines showed a correlation between the presence of the transgene and submergence tolerance. Lines overexpressing Sub1A-1 also showed increased constitutive and inducible alcohol dehydrogenase1 mRNA accumulation.

Strong expression of Sub1A-1 confers submergence tolerance to japonica rice. Elucidated mechanisms controlled by the Sub1 haplotype using rice NILs: M202, a submergence-intolerant japonica inbred grown in CA and M202(Sub1), is a descendent of a mapping population developed by hybridization of a tolerant derivative of the indica line FR13A and M202. In submerging, the tolerant M202(Sub1) displayed restrained leaf and internode elongation, chlorophyll degradation, and carbohydrate consumption, but the enzymatic activities of pyruvate decarboxylase and alcohol dehydrogenase were elevated, as compared to the intolerant M202. The transcript levels of genes associated with carbohydrate consumption, ethanolic fermentation and cell expansion were regulated in the two lines. Sub1A and 1C transcript levels were shown to be up-regulated by submergence and ethylene, with the Sub1C allele in M202 also up-regulated by treatment with gibberellic acid (GA). This demonstrated that Sub1 region haplotype determines ethylene-and GA-mediated cellular and development responses to submergence through differential expression of Sub1A and 1C. Submergence tolerance in lowland rice is conferred by a specific allele variant of Sub1A that dampens ethylene production and GA responsiveness, causing growth quiescence in growth correlated with the capacity for re-growth upon de-submergence. The project evaluated the role of Rop signal transduction in response to submergence/oxygen deprivation in the NILs. Rop-GTP levels, H_2O_2 accumulation and regulation of transcript accumulation of RopGAP (GTPase activating protein) genes in response to submergence were tested. It was observed that Sub1 genotype-specific increases in H_2O_2 and transcripts that encode three of the seven RopGAP genes, temporal accumulation of ROP-GTP was not markedly distinct in the two lines. Rop signalling is most likely upstream of Sub1 or in a distinct pathway.

6.6 Marker-assisted Backcrossing

With the success of fine mapping of Sub1 from FR13A, a sound QTL has enabled the MAB of modern rice cultivars having the ability to withstand flash flooding (Bailey-Serres et al., 2010). The polygenic QTL Sub1A was introduced to eight rice cultivars using marker-assisted backcrossing. International Rice Research Institute has successfully introgressed Sub1 QTL into high-yielding variety Swarna, which is presently adapted in many states of India (Singh et al., 2013; Dar et al., 2017). Swarna-Sub1 can endure two weeks of flash flooding and recovers well after receding of flood-water. It was released in India in 2009 after few years of successful evaluation in the farmers' fields. At the same time, Sub1 was also introgressed into mega rice varieties grown in rain-fed ecology in South and Southeast Asian countries (Neeraja et al., 2007). The Sub1 incorporated popular varieties, such as IR64-Sub1, Samba Mahsuri-Sub1, Thadokkam 1-Sub1, BR11-Sub1, BINA Dhan 11, and CR1009-Sub1 have been recognized for enhancing rice productivity in less favourable lowland ecosystem of eastern India, Nepal and Bangladesh (Ismail et al., 2013; Ahmed et al., 2016). Singh et al. (2016) also reported the transfer of Sub1 to 10 highly popular regionally-adapted Indian rice varieties. The original and Sub1 introgressed varieties show no apparent morphological variation in numerous trials in farmers' fields (Singh et al., 2017).

In a study aimed to analyze molecular diversity of submergence-tolerant genotypes of rice, 34 SSR primers were used to generate allelic variants (four alleles in RM 23843 and 27 in RM 23662). Altogether 410 allelic variants were detected at 42 SSR loci with a mean of 9.76 alleles per locus. Analysis of divergence pattern based on SSR markers allowed differentiation and classification of rice varieties into eight clusters. A large range of similarity coefficient revealed by SSR markers provided greater confidence for the assessment of genetic divergence and interrelationship among the submergence-tolerant rice genotypes. A perusal of similarity coefficients reflected a very high degree of similarity between rice genotypes 'Anh Hsung Seln' (C 11) and 'Hsung Teing', whereas 'FR 13A' and 'Swarna Sub-1' were more diverse for breeding programme to generate more recombinants. Use of SSR markers appeared more efficient in achieving unique and unambiguous characterization and differentiation of varieties used in the present study. The SSR analysis also revealed unique or variety-specific allele which could be useful as DNA fingerprints in the identification and preservation of different genotype and to develop highly submergence-tolerance varieties (Table 6.5).

Studies with a group of four indigenous and less popular rice genotypes (Meghi, Panibhasha, Jabra and Sholey) reported by growers as submergence-tolerant lines from flood-prone areas of south Bengal were explored through study of nodal anatomy, physio-biochemical screening under submergence, and genotyping with submergence-tolerance-linked rice microsatellite loci (RM loci). To identify the different allelic forms of different Sub1 components (Sub1A, Sub1B, and Sub1C) among the studied lines, the genomic DNA of individual genotypes was amplified with three ethylene response factors, like genes from Sub1 loci, located on rice chromosome 9. From the different physio-biochemical experiments performed in this investigation, it was shown that Meghi and Jabra are the two probable potent genotypes which share common properties of both submergence-tolerant and deep-water nature, whereas the rest two genotypes (Sholey and Panibhasha) behave like typical deep-water rice. The submergence-tolerance property of Meghi was also confirmed from submergence-tolerance-linked SSR-based genotyping by sharing with FR13A some common alleles as reflected in fingerprint-derived dendrogram. The rest of the genotypes shared a number of alleles and

Table 6.5: Analysis of primer pairs used for the amplification of genomic DNA extracted from 19 rice varieties.

Primers		No. of Loci	Product Size (bp)	No. of Alleles	No. of Unique Alleles	Percentage of Unique Alleles	PIC	No. of Entries having Null Alleles
RM	23668	1	102–128	11	6	54.54	0.872	0
RM	23679	1	211–253	13	8	61.53	0.908	0
RM	8303	1	129–158	12	6	50.00	0.903	0
RM	23770	1	210–361	14	11	78.57	0.903	0
RM	23778	1	274–299	6	1	16.66	0.759	0
RM	23788	1	234–283	12	7	58.33	0.903	0
RM	23805	1	225–263	9	3	33.33	0.864	0
RM	23831	1	159–194	7	2	28.57	0.814	0
RM	23843	1	284–303	4	2	50.00	0.693	11
RM	23887	1	225–260	10	5	50.00	0.858	0
RM	8300	2	191–291	19	10	52.63	0.925	0
RM	23901	2	118–299	15	13	86.66	0.927	0
RM	23902	1	268–312	12	7	58.33	0.913	1
RM	23915	1	191–251	14	10	71.42	0.914	0
RM	23917	1	230–247	6	1	16.66	0.792	0
RM	23922	1	173–248	9	3	33.33	0.880	4
RM	23928	1	153–277	14	11	78.57	0.919	0
RM	23958	2	164–270	26	20	76.92	0.949	0
RM	23996	1	253–305	11	5	45.45	0.891	0
RM	24005	1	217–269	14	10	71.42	0.914	0
RM	24103	1	105–205	12	9	75.00	0.898	3
RM	105	1	103–131	9	3	33.33	0.864	0
RM	215	3	152–542	20	8	40.00	0.922	0
RM	257	2	111–228	22	10	45.45	0.941	0
RM	316	2	166–220	11	5	45.45	0.871	0
RM	23662	2	148–315	27	20	74.07	0.956	0
RM	24013	1	153–203	8	5	62.50	0.781	0
RM	910	1	126–191	13	11	84.61	0.858	0
RM	7175	1	90–111	9	5	55.55	0.836	0
ART	5	1	220–265	10	5	50.00	0.897	0
SC	3	1	197–209	5	0	0	0.786	0
RM	296	1	112–129	7	2	28.57	0.819	0
Sub	1BC2	1	254–337	11	6	54.54	0.869	0
RM	23865	1	140–158	8	3	37.50	0.814	1

were included in a separate cluster. The common behavior of Meghi and FR13A under submergence was also confirmed from genetic study of Sub1 loci through sharing of some common alleles for three Sub1 components (Sub1A, Sub1B, and Sub1C loci). One SSR loci (RM 285) was identified as a potent molecular marker for submergence-tolerance breeding programme involving these two selected rice lines (Meghi and Jabra) as donor plants through marker-assisted selection.

To develop rice varieties with good tolerances to drought and submergence, 260 introgression lines (ILs) selected for drought tolerance (DT) from nine BC2 populations were screened for submergence tolerance (ST), resulting in 124 ILs with significantly improved ST. Results showed that genetic characterization of the 260 ILs with DNA markers identified 59 DT quantitative trait loci (QTLs) and 68 ST QTLs with an average 55% of the identified QTLs associated with both DT and ST. Approximately 50% of the DT QTLs showed 'epigenetic' segregation with very high donor introgression and/or loss of heterozygosity (LOH). Detailed comparison of the ST QTLs identified in ILs selected only for ST with ST QTLs detected in the DT-ST selected ILs of the same populations revealed three groups of QTLs underlying the relationship between DT and ST in rice: (a) QTLs with pleiotropic effects on both DT and ST; (b) QTLs with opposite effects on DT and ST; and c) QTLs with independent effects on DT and ST. Combined evidence identified most likely candidate genes for eight major QTLs affecting both DT and ST. Moreover, group b QTLs were involved in the Sub1-regulated pathway that were negatively associated with most group a QTLs (Fig. 6.7).

Submergence and drought are major limiting factors in crop production. However, very limited studies have been reported on the distinct or overlapping mechanisms of plants in response to the two water extremes. Here it was report an Ethylene Overproducer 1-like gene (OsETOL1) that modulates differentially drought and submergence tolerance in rice (*Oryza sativa* L.). Two allelic mutants of OsETOL1 showed increased resistance to drought stress at the panicle development stage. Interestingly, the mutants exhibited a significantly slower growth rate under submergence stress at both the seedling and panicle development stages. Over-expression (OE) of OsETOL1 in rice resulted in reverse phenotypes when compared with the mutants. The OsETOL1 transcript was differentially responsive to abiotic stresses. OsETOL1 was found to interact with OsACS2, a homolog of 1-amino-cyclopropane-1-carboxylate (ACC) synthase (ACS), which acts as a rate-limiting enzyme for ethylene biosynthesis. In the osacs2 mutant and OsETOL1-OE plants, ACC and ethylene content decreased significantly, and exogenous ACC restored the phenotype of osetol1 and OsETOL1-OE to wild-type under submergence stress, implying a negative role for OsETOL1 in ethylene biosynthesis. The expression of several genes related to carbohydrate catabolism and fermentation showed significant changes in the osetol1 and OsETOL1-OE plants, implying that OsETOL1 may affect energy metabolism. These results together suggest that OsETOL1 plays distinct roles in drought and submergence tolerance by modulating ethylene production and energy

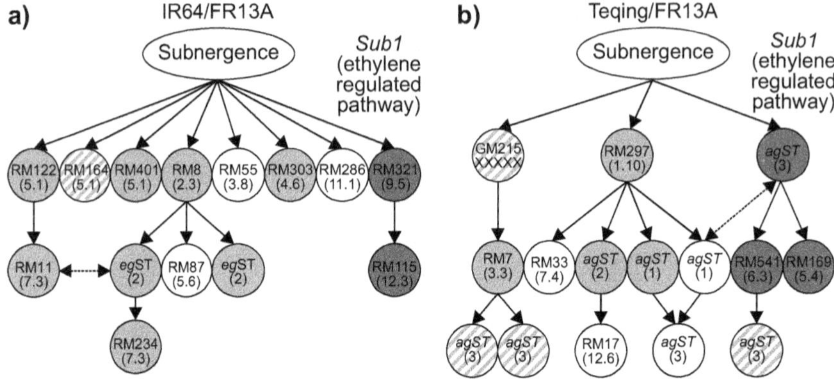

Fig. 6.7: Genetic overlap between drought tolerance (DT) and submergence tolerance (ST) revealed in genetic networks (multi-locus structures) constructed from ST loci from the two BC2F2 populations, in which each oval represents a locus or association group (AG) identified in 14 BC2F2 plants selected under submergence stress from population IR64/FR13A//IR64 (A) and from eight BC2F2 plants of population Teqing/FR13A//Teqing (B) (Wang et al., 2015). Those loci in green color were detected as ST loci identified in the 10 BC2F4 lines selected first for DT and then for ST from population A and in 13 BC2F4 lines selected for DT-ST from population B (Table S3), respectively, while those ovals in patched green represent loci detected with closely linked markers of the same bins. Those ovals with red borderlines were DT loci identified in the same populations, while those ovals in blue and while colours were not detected in the DT-ST lines in the two populations.

metabolism. Findings from the expression and functional comparison of three ethylene overproducer (ETOL) family members in rice further supported the specific role of OsETOL1 in the responses to the two water stresses (Fig. 6.8 and 6.9).

Swarna (MTU7029) is a mega-variety from India, and it is currently grown on more than 6 million ha in rain-fed lowland areas of South Asia. It has high yield potential and preferred quality characteristics but is intolerant of most abiotic stresses. The introgression of the *Sub1* QTL into Swarna and a few other varieties through MABC successfully improved their submergence tolerance under field conditions (Xu et al., 2006; Neeraja et al., 2007; Septiningsih et al., 2009), with a yield advantage of 2–3.5 t ha^{-1} under natural field conditions (Singh et al., 2009).

Singh et al. (2011) observed that the QTL is not effective in semi-dwarf rice cultivars under partial SF; cultivars like Swarna and SwarnaSub1 were more sensitive to long-term partial SF than inherently taller cultivars. Earlier it was reported that there was a greater loss of yield under submergence in SwarnaSub1 compared to traditional cultivar Balidhan61 and significant variation among cultivars in stem elongation ability under submergence (Sarkar and Bhattacharjee, 2011). These reports denote that resilience of genotypes to SF depends less on Sub1-introgression and more so on the genetic make-up. This is in conformity with observation of

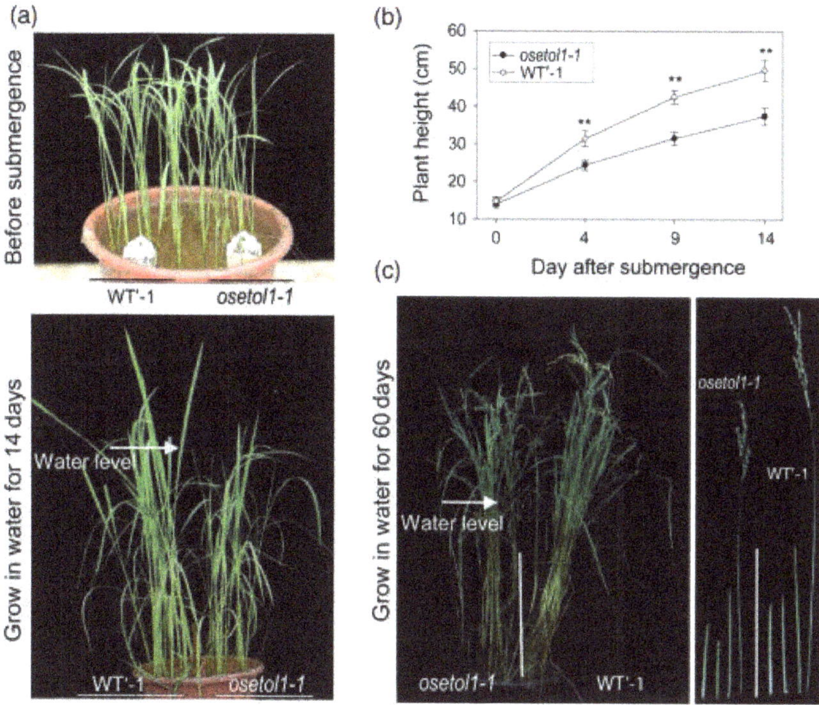

Fig. 6.8: Phenotype of the osetol1 mutants under submergence. (a, b) The osetol1 mutant showed significantly increased sensitivity to partial submergence (complete submergence for three days and then the water level was maintained so that the top leaf tips were exposed, as indicated by the white arrow, for 14 days) at the seedling stage, as indicated by the reduced growth rate. Asterisks indicate significant difference (t-test), $^{**}P < 0.01$ level, values are means deviation (SD) (n = 8). (c) The osetol1 mutant showed significantly increased sensitivity to partial submergence (complete submergence for three days and then the water level was maintained, as indicated by the white arrow, for up to 60 days) during the vegetative and reproductive stages, as indicated by the reduced plant height.

Singh et al. (2014) that the physiological basis of tolerance to submergence in rice involves genetic factors in addition to Sub1 gene. Both Savitri and Swarna are semi-dwarf cultivars, and they produced more ethylene for greater elongation of stem under SF irrespective of Sub1-introgression. Additionally, Sub1 might act in a different manner under flash flooding and SF. The Sub1-introgressed cultivars of the study were under SF over a long period from vegetative to grain-filling and their response was not identical to plants encountering other types of flooding. Further, the expression mode of Sub1 in vegetative tissues in restricting the elongation growth, may not have a similar response in the reproductive tissues. It is concluded that under SF, semi-dwarf Sub1 cultivars are less suitable for cultivation than the corresponding parental lines. However, this inference reached for the mega semi-dwarf lines, used in the study, may not hold

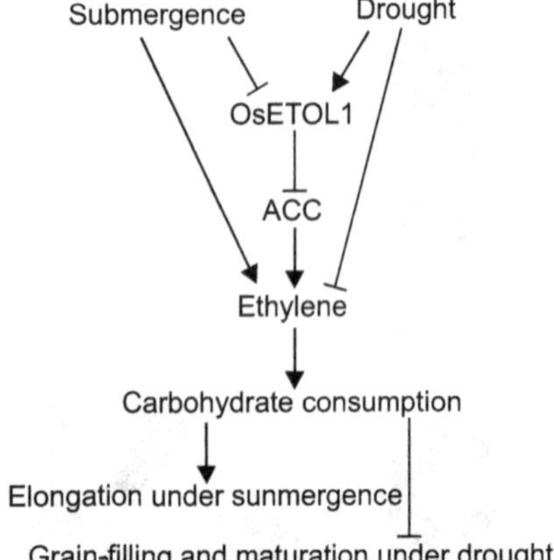

Fig. 6.9: Working model for the function of OsETOL1 in response to drought and submergence stresses.

true for the cultivars with taller genetic backgrounds that would have different effects under SF. Future studies should investigate the possible differential response on other type of Sub1 cultivars (the ones that have moderate elongation rate under SF and/ or Sub1 cultivar with inherently tall attribute) to see their suitability of planting in areas that are affected by the combination of flash flooding and SF stress.

In recent times, Submergence1 (Sub1) QTL, responsible for imparting tolerance to flash flooding, has been introduced in many rice cultivars, but resilience of the QTL to stagnant flooding (SF) is not known. The response of Sub1-introgression has been tested on physiology, molecular biology and yield of two popular rice cultivars (Swarna and Savitri) by comparison of the parental and Sub1-introgression lines (SwarnaSub1 and SavitriSub1) under SF. Compared to control condition SF reduced grain yield and tiller number but increased plant height and Sub1—introgression mostly matched these effects. SF increased ethylene production by over-expression of ACC-synthase and ACC-oxidase enzyme genes of panicle before anthesis in the parental lines. Expression of the genes changed with Sub1-introgression, where some enzyme isoform genes over-expressed after anthesis under SF. Activities of endosperm starch-synthesizing enzymes SUS and AGPase declined concomitantly with rise in ethylene production in the Sub1-introgressed lines, resulting in low starch synthesis and accumulation of soluble carbohydrates in the developing spikelets. In conclusion, Sub1-introgression into the cultivars increased susceptibility to SF. Subjected to SF, the QTL promoted genesis of ethylene in the panicle at anthesis to the detriment of grain yield, while compromising with morphological features, like tiller production and stem elongation (Fig. 6.10).

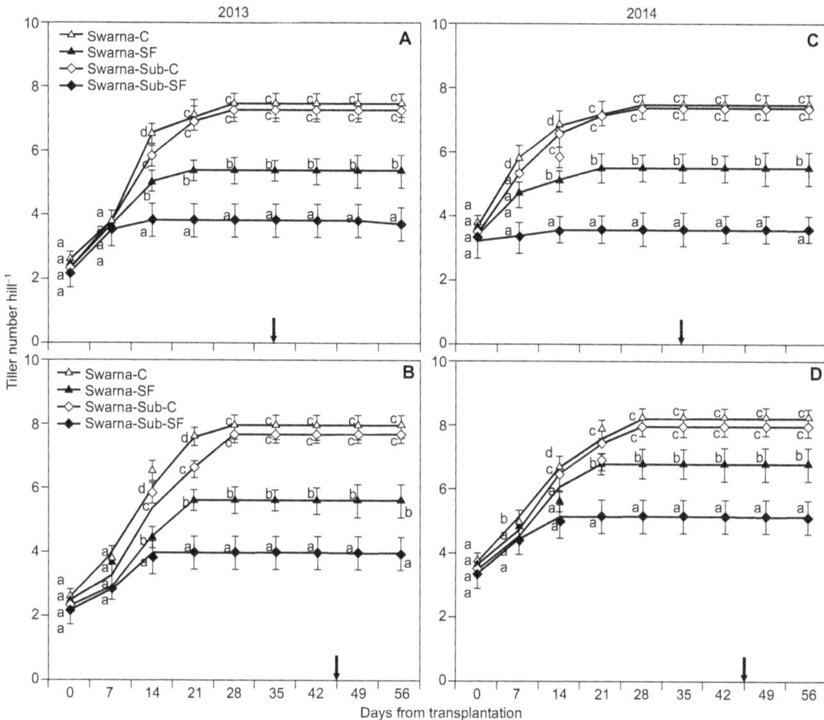

Fig. 6.10: Tillering in Swarna and Savitri and sub1types grown on 2013 and 2014.

Average grain weight of panicle of main shoot increased slowly in a sigmoidal pattern during post-anthesis period in all cultivars. The temporal increase of grain weight was greater in control compared to SF. Variation in cultivar, flooding treatment and days after anthesis significantly ($p^{**} < 0.01$) influenced the grain weight gain in both the years. Between the cultivars, Savitri had larger grain weight compared to Swarna. SF impacted grain weight significantly during the post-anthesis period and extended the period of maturation in all cultivars. Introgression of the Sub1 in Swarna reduced grain weight gain, but it was not much effective for Savitri. SavitriSub1 possessed better grain growth with passage of time as against its parent Savitri. Rate of grain-filling in the panicle of the main shoot exhibited a curvilinear pattern temporally in the post-anthesis period in all cultivars. The rate increased up to day 13 after anthesis and declined thereafter, till maturity in plants under control condition. In comparison, under SF, grain-filling rate increased slowly to day 16 and declined continuously thereafter to day 28. In the F-test, cultivar, treatment and days after anthesis and their interaction significantly ($^{***}p < 0.001$) influenced the grain-filling rate, but impact of the SF treatment was wider than the others. In control condition, grain-filling rate was almost similar in Swarna and SwarnaSub1. Under SF, however, grain-filling rate was greater most of the time in Swarna compared to SwarnaSub1. The trend was somewhat different between Savitri and SavitriSub1. The grain-filling rate was greater in SavitriSub1 as compared to Savitri, both in control and SF conditions mostly at the middle stage of grain growth. Grain growth rate ceased at least three days early in control compared to SF condition.

6.7 Multiple Stress-tolerant Varieties

Flash flooding (FF) along with bacterial blight (BB) outbreak are very destructive for rice production in the rain-fed shallow-lowland (RSL) ecosystem. The presence of dynamic Xoo races with varying levels of genetic diversity and virulence renders their management extremely challenging under RSL. In this context, the marker-assisted improvement of plant resistance/tolerance has been proven as one of the most promising strategies towards the development of sustainable cultivars. The present study demonstrates the marker-assisted introgression of the submergence tolerant (Sub1) and three bacterial blight-resistant genes (Xa21 + xa13 + xa5) into the genetic background of Hasanta, a long duration popular rice variety in the eastern coastal region of India. The rice genotypes, Swarna Sub1 (carrying Sub1) and IRBB66 (carrying Xa21 + xa13 + Xa7 + xa5 + Xa4) had maximum genetic similarity (0.84 and 0.73, respectively) with Hasanta; recurrent parent (RP) was used as the donor. The forward analysis of target genes in F1s, IC1F1s, and backcross (BC) generations was performed by linked/genic markers (Sub1bc2; pTA248, xa13prom and RM122), whereas background recovery of RP in each BC and segregating generations was performed by utilizing 108 hypervariable SSR markers. Intervened speed breeding (SB) strategy and intensive phenotyping could lead the development of near isogenic lines (NILs) as to the RP in all basic traits. The performance of the near isogenic lines (NILs, BC2F3 and BC2F4), HS 232-411-391-756-37, HS 232-411-391-809-8, HS 232-411-391-756-18, HS 110-224-197-10-36, HS 232-411-391-809-81, HS 110-224-197-10-41, and HS 232-411-391-809-63 establishes the utility of marker-assisted backcross-breeding (MAB) and SB in accelerated trait introgression. The introgressed lines carrying Sub1 + Xa21 + xa13 + xa5 showed 76% to 91% survival under 14 days of submergence and durable BB resistance (per cent disease index-PDI of 2.68 ± 0.26 to 6.22 ± 1.08 and lesion length, LL of 1.29 ± 0.12 to 4.2 ± 0.64 cm). Physiological analysis revealed that improved NILs, carrying Sub1 gene conquered adaptive physiological modulations, reduced the consumption of soluble sugar and the degradation of total chlorophyll contents (TCC), and an enhanced level of alcohol dehydrogenase activity (ADH) and proline accumulation in all submergence regimes (Table 6.6). The pyramided lines attained complete product profile of RP, that will

Table 6.6: ANOVA of physiological traits in parents and NILs (Hasanta/Swarna Sub1// Hasanta/ IRBB66) under submergence (0, 7, 14, 21 days) condition

Source	DF	Mean Sum of Squares			
		ADH	TSS	PA	TCC
Replications	2	39.017*	0.00019*	0.0023*	2.424*
Genotype (A)	14	269.125**	0.249**	0.165**	4.234**
Treatment (B)	3	26.491**	0.0622**	0.0226**	2.571**
AxB	42	22.230**	0.172**	0.0422**	3.121**
Error	42				

Error 42 Note: *, ** significance at 5% and 1% probability level, ADH, Alcohol dehydrogenase; TSS, total soluble sugar; PA, proline accumulation; TCC, total chlorophyll content.

contribute to sustainable rice production under RSL, particularly in the coastal region that has substantial acreage under the variety Hasanta (Das et al., 2022).

6.8 Mega Varieties

The introgression of the Sub1 region through marker-assisted breeding has significantly enhanced submergence tolerance in several mega-varieties without limiting their development, yield, or grain quality (Singh et al., 2009; Sarkar and Panda, 2009). These new lines are capable of enduring submergence for the duration of flooding after the seedling stage but before flowering, and the flood completely subsides within 10 to 20 days, depending on flood-water conditions (Das et al., 2009), although in some Sub1 varieties, vegetative growth did not occur until the water level declined to 10–15 cm. This was primarily due to the short stature of these varieties. Recent studies have shown that this phenomenon was not observed when Sub1 was incorporated into taller varieties or those with the ability to tolerate partial stagnant flooding (20–50 cm). The yield improvement occurring due to the Sub1 introgression in the lines was believed to greatly stabilize their yield in rain-fed lowland environments characterized by flash flooding. Contrastingly, deep-water rice, unlike the Sub1 rice, escapes stagnant partial flooding by enhancing the growth of internodes (Voesenek and Bailey-Serres, 2015). Due to the difference through which deep-water rice genes SK1/SK2 and the submergence-tolerance gene Sub1A regulate ethylene-mediated GA responsiveness, it is unlikely that these genes can be combined to develop genotypes that are tolerant to both stagnant and submergence flooding. Advances in knowledge of the regulatory networks vital for balancing energy management and growth under submerged conditions can assist in identifying natural genetic diversity in some loci, which could be exploited to enhance the adaptability of rice to submergence. The reassessment of QTLs from FR13A and other varieties lacking Sub1A-1 but moderately tolerant to flooding can be useful in further improving submergence tolerance. There is a need to increase the yield of rice in marginal growing regions. This, among other things, can be achieved by generating rice genotypes which combine submergence tolerance with tolerance to other environmental stress conditions and characteristic grain quality that is preferred in the local market. Research has shown that early germinating rice varieties with early coleoptile elongation in anaerobic soil have been recognized and used in breeding programs at IRRI (Angaji et al., 2010; Singh et al., 2017) for direct seeding (Table 6.7). The expression of the mentioned traits in specific developmental windows can enable its incorporation alongside Sub1. In coastal areas prone to flooding as a result of tidal waves, it would be beneficial to have rice that can tolerate saline flood-water. Also, in non-irrigated lowland regions where rainfall is the only source of water, growers will benefit from rice which is capable of tolerating drought. The success in combining submergence tolerance with other qualities may promote the recognition of other landraces, for which traits will be unearthed at molecular level before getting to the grower's field as improved varieties. It is believed that with the advancement in plant breeding and innovations in rice germplasm collection, these challenges will be overcome in due time to meet the increasing global demand for food.

Excessive water stresses, such as waterlogging and submergence, adversely affect rice growth and grain yield. While some historical rice cultivars exhibit notable resilience to submergence, their total yield suffers (Singh et al., 2014). In contrast, modern rice cultivars are sensitive to flooding, often resulting in farmers losing their whole crop. Rice plants can perish soon after flooding due to high energy expenditure and protein hydrolysis during submergence. Flooding degrades the quality of endosperm reserves, adversely affecting the nutritional value and milling and cooking properties of rice grain (Zhou et al., 2020). Genomic techniques have been used to investigate how abiotic stresses affect grain development (Verma et al., 2021), with several genetic regulators of tolerance identified and successfully used to improve rice cultivars. For example, genetic loci controlling salinity stress have been discovered and pyramided to develop green super rice types (Pang et al., 2017). Using marker-assisted breeding, Kumar et al. (2018) combined quantitative trait loci (QTL) for submergence and drought tolerance to identify varieties with high yield potential, validating their performance by exposing them to various stresses.

Submergence reduces the quality and quantity of rice, especially when it occurs during the reproductive and maturity stages. Submergence

Table 6.7: Genetic loci controlling anaerobic germination and high seedling survival under submergence.

Trait	QTLs/Gene	Chromosome	Flanking Markers	References
High survival	qSub1.1	1	id1000556-id1003559	Gonzaga et al., 2016
High survival	qSub4.1	4	Id4010261-id4012434	"
High survival	qSub8.1	8	Id08005815-id08007472	"'
High survival	qSub10.1	10	Id10005538-RM25835	"'
Anaerobic Germination	qAG-5	5	RM405-RM249	Jiang et al., 2006
Anaerobic Germination	qAG-7-2	7	RM21868-RM172 Seq-rs3583	Angaji et al., 2010; Zhang et al., 2017
Anaerobic Germination	qAG-7-1,AG2	7	RM3583-RM21427	Septiningsih et al., 2013
Anaerobic Germination	qAG-9-2, AG1	9	RM3769-RM105 Seq-rs4216	Angaji et al., 2010; Zhang et al., 2017
Anaerobic Germination	qAG-11	11	RM21-RM22 Seq-rs5125	""
Anaerobic Germination	qAG-1-2	1	RM11125-RM104 Id29187939-Id32847451	""
Anaerobic Germination	qAG-3	3	RM7097-RM520	Angaji et al., 2010
Anaerobic Germination	qAG-9-1	9	RM8303-RM5526	""
High Survival	qSub.1	8	8,608,433-8,686,009	Gonzaga et al., 2017
HighSurvival	qSub2.1	2	2,430,179-2,470,790	"'

significantly delays flowering and maturity, reducing grain yield, shoot biomass, harvest index, and yield components (Marndi et al., 2022). Reductions in grain filling, grain number per panicle, and grain weight are primarily responsible for decreased grain production due to submergence (Kato et al., 2014). Submergence during the vegetative stage affects critical grain quality parameters, with a higher proportion of hull, brown rice, and bran in rough rice compared to non-stressed counterparts, as well as chalky grains, breakage during hulling, and reduced proportion of amylose, but increased crude protein content.

Yield losses due to submergence are attributable to a smaller sink size/capacity and reduced carbohydrate metabolism and thus reduced partitioning into grain. Djali et al. (2012) reported that submerged rice had higher protein, moisture, and amylase contents than the control plants but lower yield, hardness, stickiness, and brightness.

The changing climate and resultant rise in sea-water levels lead to unexpected spells of multiple abiotic stresses at different stages of paddy production. In coastal areas, increasing temperatures, erratic rainfall, and inundation of saline water due to sea-level rise can change the micro-environment in fields. Studies are limited in this arena for rice. Tolerant rice genotypes adapt to combined salinity and submergence due to the presence of well-developed constitutive aerenchyma and increased ethylene production and respiratory burst oxidase homolog (RBOH) signalling. RBOH-mediated ROS production resulted in the development of constitutive aerenchyma in a saline- and flooding-tolerant rice variety, Rashpanjor (Chakraborty et al., 2021). Chlorophyll fluorescence imaging identified tolerant varieties under combined salinity and partial submergence (Pradhan et al., 2018).

Drought has had a significant impact on rice cultivation and production in Malaysia. Using the marker-assisted breeding technique, the current study was conducted to introgress the submergence tolerance QTL (*Sub1*) from IR64-*Sub1* into UKM91 with drought yield QTL ($qDTY_{3.1}$). The effects of *Sub1* introgression on agro-morpho-physiological traits under reproductive stage drought stress (RSD) and non-stress (NS) conditions, as well as survival percentage under vegetative stage submergence stress (VS) are reported in a study. In both RSD and NS, the effect of *Sub1* presence alone outperformed the effect of *Sub1* + $qDTY_{3.1}$ combination for grain yield (GY). Under RSD, lines with only *Sub1* had a higher number of panicles and thousand-grain weight than lines with *Sub1* + $qDTY_{3.1}$ and $qDTY_{3.1}$ only indicating that *Sub1* plays multiple roles in conferring drought tolerance. In RSD and NS, the best-performing lines achieved GY advantages of up to 1713 kg/ha and 1028 kg/ha over UKM91 respectively. After 14 days of VS, the selected lines from NS and RSD trials (GEN11, GEN19, GEN37, GEN49, GEN66, and GEN72) had a 100% survival rate while IR64-*Sub1* had an 86.67% survival rate due to reduction in chlorophyll content and a low shoot elongation percentage. The developed lines have the potential to mitigate the effects of climate change and reduce yield loss while also serving as pre-breeding materials for breeding programmes.

Other unidentified genetic factors for submergence tolerance are expected to be used as new tools for further improvement of molecular breeding. Since wild rice has different mechanisms from cultivated rice, at least regarding flash flood tolerance (Niroula et al., 2012; Okishio et al., 2015), it will be possible to apply these specific factors to further molecular breeding. Introgressions by classical backcrossing or marker-assisted backcrossing have been applied in the development of useful rice varieties. Recently, genome-editing technologies using engineered site-specific nucleases, particularly the CRISPR/Cas9 system, have been advanced drastically (Doudna and Charpentier, 2014). These new technologies facilitate the functional modification of genes related to useful crop traits by producing site-directed point mutations, deletions, or insertions (Mishra et al., 2018). The identified genetic factors, such as Sub1A-1, OsTPP7, SK1/2, and SD1-DW can be applied to genome editing for improved submergence tolerance. In addition, application of genome editing to genes for anatomical adaptation to excess water (i.e., genes regulating the formation of aerenchyma, ROL, and leaf gas film) may help the breeding of flooding-tolerant rice, which is impossible by conventional breeding.

Rice XLGs play vital roles in key agronomic traits. XLGs are important in vegetative growth, panicle and seed development, and tolerance of both abiotic and biotic stresses. There are numerous inconsistencies in the reported mutant phenotypes and stress responses of rice XLG CRISPR alleles. It will be important to determine whether these apparent discrepancies arise from the cultivar background and/or the specific growth conditions. It is also important to evaluate whether the CRISPR alleles produced in each study are true nulls. Even though the predicted protein structures of these CRISPR alleles suggest a loss of functionality, particularly for frame-shift and stop-mutations, the possible production of a sequence-altered protein or a truncated XLG that yields anomalous results cannot be ruled out without further investigation. Another key unanswered question is possible functional partitioning or synergism between the canonical Gα RGA1 and the rice XLGs in plant architecture and development, as has been observed for the maize Gα subunits. It also will be crucial to determine specific functions of the N-terminal domain that could provide possible insights regarding XLG signalling.

References

Adkin, S.W., Shiraishi, T., and McComb, J.A. (1990). Submergence tolerance of rice – a new glasshouse method for the experimental submergence of plants. *Physiologia Plantarum*, 80: 642–646.

Ahmed, S., Ahmad, M., Swami, B.L., and Ikram, S. (2016). A review on plants extract mediated synthesis of silver nanoparticles for antimicrobial applications: A green expertise. *J. Adv. Res.*, 7: 17–28.

Angaji, S.A., Septiningsih, E.M., Mackill, D.J., and Ismail, A.M. (2010). QTLs associated with tolerance of flooding during germination in rice (*Oryza sativa* L.). *Euphytica*, 172: 159–167.

Angaji, S.A. (2008). Mapping QTLs for submergence tolerance during germination in rice. *Afa. J. Biotechnol.*, 7: 2551–8.

Asano, K., Yamasaki, M., Takuno, S., Miura, K., Katagiri, S., Ito, T. et al. (2011). Artificial selection for a green revolution gene during japonica rice domestication. *Proc. Natl. Acad. Sci.* USA, 108: 11034–11039.

Bailey-Serres, J., Fukao, T. Ronald, P.C., Ismail, A.M., Heuer, S., and Mackill, D.J. (2010). Submergence tolerant rice: SUB1's journey from landrace to modern cultivar. *Rice,* 3: 138–147.

Baltazar, M.D., Ignacio, J.C.I., Thomson, M.J., Ismail, A.M., Mendioro, M.S., and Septiningsih, E.M. (2014). QTL mapping for tolerance of anaerobic germination from IR64 and the aus landrace Nanhi using SNP genotyping. *Euphytica,* 197: 251–60.

Baltazar, M.D., Ignacio, J.C.I., Thomson, M.J., Ismail, A.M., Mendioro, M.S., and Septiningsih, E.M. (2014). QTL mapping for tolerance of anaerobic germination from IR64 and the aus landrace Nanhi using SNP genotyping. *Euphytica,* 197: 251–260.

Chen, S., Wang, J., Pan, Y., Ma, J., Zhang, J., Zhang, H. et al. (2012). Genetic analysis of seed germinability under submergence in rice, *Chin. Bull. Bot.,* 47: 28–35.

Chakraborty, K., Ray, S., Vijayan, J., Molla, K., Nagar, R., Jene, P., et al. (2021). Preformed aerenchyma determines the differential tolerance response under partial submergence imposed by fresh and saline water flooding in rice, *Physiologia Plantarum,*173(4). Doi: 10.1111/ppl.13536.

Dar, M.H., Chakraborty, R., Waza, S.A., Sharma, M., Jaidi, N.W., Singh, A.N. et al. (2017). Transforming rice cultivation inflood-pronecoastal odisha to ensure food and economic security. *Food Sec.,* 9: 711–722.

Das, G., Pradhan, B., Bastia, D., Samantaray, S., Jena, D., Rout, D. et al. (2022). Pyramiding Submergence Tolerance and Three Bacterial Blight Resistance Genes in Popular Rice Variety Hasanta through Marker-assisted Backcross Breeding, *Agriculture,* 12(11): 1815.

Das, K.K., Panda, D., Sarkar, R.K., Reddy, J.N., and Ismail, A.M. (2009). Submergence tolerance in relation to variable floodwater conditions in rice, *Environ. Exp,. Bot., 66*: 425–434.

Djali, M., Nurhasanah, S., Lembong, E., and Rahmat, R. (2012). The effect of flooding on rice characteristics of *Cilamaya muncul'* on various days after planting during the last reproductive and maturation phase, II *Asia Pacific Symposium Postharvest Res. Educ. Extension*: *APS2012,* 1011: 285–291.

Doudna, J.A., and Charpentier, E. (2014). Genome editing. The new frontier of genome engineering with CRISPR-Cas9, *Science,* **346**(6213): 1258096.

Eiguchi, M., Hirano, H.Y., Sano, Y., and Morishima, H. (1993). Effect of water depth on internode elongation and floral induction in a deep-water–tolerant rice line carrying the dw3 gene, *Jpn. J. Breed.,* 431: 35–139.

Faris, J.D., Li, W.L., Liu, D.J., Chen, P.D., and Gill, B.S. (1999). Candidate gene analysis of quantitative disease resistance in wheat, *Theor. Appl. Genet.,* 98: 219–225.

Fukao, T., Xu, K., Ronald, P., Bailey-Serres, J., and Fukao, T. (2006). A variable cluster of ethylene response factor like genes regulates metabolic and developmental acclimation responses to submergence in rice. *Plant Cell,* 18: 2021–2034.

Fukao, T., and Bailey-Serres, J. (2008). Submergence tolerance conferred by SUB1A is mediated by SLR1 andSlRl1 restriction of gibberellin responses in rice. *Proc. Natl. Acad. Sci.* USA, 105: 16814–16819.

Fukao, T., Harris, T., and Bailey-Serres, J. (2009). Evolutionary analysis of the Sub1 gene cluster that confers submergence tolerance to domesticated rice, *Ann. Bot.,* 103: 143–150.

Gibbs, D.J., Lee, S.C., Isa, N.M., Gramuglia, S., Fukao, T., Bassel, G.W. et al. (2011). Homeostatic response to hypoxia is regulated by the N-end rule pathway in plants, *Nature,* 479(7373): 415–418.

Gill, M.B., Zeng, F., Shabala, L., Zhang, G., Fan, Y., Shabala, S. et al. (2017). Cell-based phenotyping reveals QTL for membrane potential maintenance associated with hypoxia and salinity stress tolerance in barley, *Front. Plant Sci., 8*: 1941.

Gomez-Ariza, J., Brambilla, V., Vicentini, G., Landini, M., Cerise, M., Carrera, E. et al. (2019). A transcription factor coordinating internode elongation and photoperiodic signals in rice, *Nat. Plants,* 5(4): 358–362.

Gonzaga, Z.J.C., Carandang, J., Sanchez, D.L., Mackill, D.J., and Septiningsih, E.M. (2016). Mapping additional QTLs from FR13A to increase submergence tolerance beyond Sub1, *Euphytica*, 209: 627–636.

Gonzaga, Z.J.C., Carandang, J., Singh, A., Collard, B.C.Y., Thomson, M.J., and Septiningsih, E.M. (2017). MappingQTLs for submergence tolerance in rice using a population fixed for SUB1A tolerant allele. *Mol. Breed.*, 37: 47–53.

Guglielminetti, L., Yamaguchi, J., Perata, P., and Alpi, A. (1995). Amylolytic Activities in Cereal Seeds under Aerobic and Anaerobic Conditions, *Plant Physiol.*, 109: 1069–1076.

Hamamura, K., and Kupkanchankul, T. (1979). Inheritance of floating ability of rice, *Jpn. J. Breed.*, *29*: 211–216.

Haque, Q.A., Hille Ris Lambers, D., Tepora, N.M., and dela Cruz, Q.D. (1989). Inheritance of submergence tolerance in rice, *Euphytica*, 41: 247-251.

Hattori, Y., Miura, K., Asano, K., Yamamoto, E., Mori, H., and Kitano, H. (2007). A major QTL confers rapid internode elongation in response to water rise in deep-water rice, *Breed. Sci.*, 57: 305–314.

Hattori, Y., Nagai, K., Mori, H., Kitano, H., Matsuoka, M., and Ashikari, M. (2008). Mapping of three QTLs that regulate internode elongation in deep–water rice, *Breed. Sci.*, 58: 39–46.

Hattori, Y., Nagai, K., Furukawa, S., Xian-Jun, S., Ritsuko, K., Hitoshi, S. et al. (2009). The ethylene response factors SNORKEL1 and SNORKEL2 allow rice to adapt to deep water. *Nature*, 460: 1026–1030.

Hsu, S., and Tung, C. (2015). Genetic mapping of anaerobic germination-associated QTLs controlling coleoptile elongation in rice, *Rice*, 8: 38.

Huang, N., Sutliff, T.D., Litts, J.C., and Rodriguez, R.L. (1990). Classification and characterization of the rice alpha-amylase multigene family, *Plant Mol. Biol.*, 14: 655–668.

Huang, N., Stebbins, G., and Rodriguez, R. (1992). Classification and evolution of α-amylase genes in plants, *Proc. Natl. Acad. Sci.* USA, 89: 7526–7530.

Hwang, Y.S., Thomas, B.R., and Rodriguez, R.L. (1999). Differential expression of rice α-amylase genes during seedling development under anoxia, *Plant Mol. Biol.*, 40: 911–920.

Ismail, A.M., Ella, E.S., Vergara, G.V., and Mackill, D.J. (2009). Mechanisms associated with tolerance to flooding during germination and early seedling growth in rice (*Oryza sativa*), *Ann Bot.*, 103: 197–209.

Ismail, A.M., Singh, U.S., Sudhanshu Singh, Dar, M.H., and Mackill, D.J. (2013). The contribution of submergence-tolerant (Sub1) rice varieties to food security in flood -prone rainfed lowland areas in Asia. *Field Crops Res.*, 152: 83–93.

Jackson, M.B., Waters, I., Setter, T., and Greenway H. (1987). Injury to rice plants caused by complete submergence: A contribution by ethylene (ethene), *Journal of Experimental Botany*, 38: 1826-1838.

Jiang, L., Hou, M.Y., Wang, C.M., and Wan, J.M. (2004). Quantitative trait loci and epistatic analysis of seed anoxia germinability in rice (*Oryza sativa*), *Rice Sci.*, 11: 238–44.

Jiang, L., Liu, S., Hou, M., Tang, J., Chen, L., Zhai, H. et al. (2006). Analysis of QTLs for seed low temperature germinability and anoxia germinability in rice (*Oryza sativa* L.), *Field Crop Res.*, 98: 68–75.

Jung, K.H., Seo, Y.S., Walia, H., Cao, P., Fukao, T., Canlas, P.E. et al. (2010). The submergence tolerance regulator Sub1A mediates stress-responsive expression of AP2/ERF transcription factors, *Plant Physiol.*, 152(3): 1674–1692.

Kato, Y., Collard, B.C., Septiningsih, E.M., and Ismail, A.M. (2014). Physiological analyses of traits associated with tolerance of long-term partial submergence in rice, *AoB Plants*, 6: plu058.

Kawano, R., Doi, K., Yasui, H., Mochizuki, T., and Yoshimura, A. (2008). Mapping of QTL for floating ability in rice, *Breed. Sci.*, 58: 47–53.

Kim, S., and Reinke, R.F. (2018). Identification of QTLs for tolerance to hypoxia during germination in rice, *Euphytica*, 214: 160.

Kretzschmar, T., Pelayo, M.A.F., Trijatmiko, K.R., Gabunada, L.F., Alam, R., and Jimenez, R. (2015). A trehalose-6-phosphate phosphatase enhances anaerobic germination tolerance in rice, *Nat. Plants*, 1: 15124.

Kumar, A., Sandhu, N., Dixit, S., Yadav, S., Swamy, B.P.M., and Shamsudin, N.A.A. (2018). Marker-assisted selection strategy to pyramid two or more QTLs for quantitative trait-grain yield under drought, *Rice*, 11: 35. Doi: 10.1186/s12284-018- 0227-0

Kuroha, T., Nagai, K., Gamuyao, R., Wang, D.R., Furuta, T., Nakamori, M., et al. (2018). Ethylene-gibberellin signalling underlines adaptation of rice to periodic flooding. *Science*, 361: 181–186.

Lin, C.C., Chao, Y.T., Chen, W.C., Ho, H.Y., Chou, M.Y., Li, Y.R. et al. (2019). Regulatory cascade involving transcriptional and N-end rule pathways in rice under submergence, *Proc. Natl. Acad. Sci.*, U S A, **116**(8): 3300–3309.

Locke, A.M., Barding, G.A. Jr, Sathnur, S., Larive, C.K., and Bailey-Serres, J. (2018). Rice Sub1A constrains remodelling of the transcriptome and metabolome during submergence to facilitate post-submergence recovery, *Plant Cell Environ.*, **41**(4): 721–736.

Loreti, E., Valeri, M.C., Novi, G., and Perata, P. (2018). Gene regulation and survival under hypoxia requires starch availability and metabolism. *Plant Physiol.*, 176: 1286–1298.

Mackill, D.J. (2006) Breeding for resistance to abiotic stresses in rice: The value of quantitative trait loci. In: Lamkey, K.R., Lee, M. (Eds.). Plant Breeding: The Arnel R. Hallauer International Symposium. Blackwell Publication: Ames, IA, U S A, pp. 201–212. [Google Scholar]

Mahender, A., Anandan, A., and Pradhan, S.K. (2015). Early seedling vigour, an imperative trait for direct-seeded rice: an overview on physio-morphological parameters and molecular markers, *Planta*, 241: 1027–50.

Manangil, O.E., Vu, H.T.T., Mori, N., Yoshida, S., and Nakamura, C. (2013). Mapping of quantitative trait loci controlling seedling vigor in rice (*Oryza sativa* L.) under submergence, *Euphytica*,192: 63–75.

Marndi, B.C., Anilkumar, C., Muhammed, Azharudheen, T.P., Sah, R.P., Moharana, D., et al. (2022). Cataloguing of rice varieties of NRRI suitable for different abiotic stress-prone ecologies. Climate resilient technologies for rice-based production systems in eastern India. ICAR National Rice Research Institute, Bhattacharyya, P., Chakraborty, K., Molla, K.A., Poonam, A., Bhaduri, D., Sah, R.P., et al. (Eds.). India: Cuttack, Odisha, 408.

Mishra, S.S., Behera, P.K., Kumar, V., Lenka, S.K., and Panda, D. (2018). Physiological characterization and allelic diversity of selected drought tolerant traditional rice (*Oryza sativa* L.) landrace of koraput, India. *Physiol. Mol. Biol. Plants*, 24: 1036–1046.

Mohanty, H.K., Suprihatno, B., Khush, G.S., Coffman, W.R., and Vergara, B.S. (1981). Inheritance of submergence tolerance in deepwater rice. Proceds. *The 1981 International Deepwater Rice Workshop*. pp. 121–134.

Mohanty, H.K., and Khush, G.S. (1985). Diallel analysis of submergence tolerance in rice (*Oryza sativa* L.), *Theoretical and Applied Genetics*, 70: 467-473.

Nagai, K., Kuroha, T., Ayano, M., Kurokawa, Y., Angeles-Shim, R.B., and Shim, J.H. (2012). Two novel QTLs regulate internode elongation in deep-water rice during the early vegetative stage, *Breed. Sci.*, 62: 178–185.

Nagai, K., Kondo, Y., Kitaoka, J., Noda, T., Kuroha, T., Angeles-Shim, R.B. et al. (2014). QTL analysis of internodal elongation in response to gibberellin in deep water rice. *AoB Plants*, 6: plu028.

Nakano, T., Suzuki, K., Fujimura, T., and Shinshi, H. (2006). Genome-wide analysis of the ERF gene family in Arabidopsis and rice. *Plant Physiol.*, 140: 411–432.

Nandi, S., Subudhi, P.K., Senadhira, D., Manigbas, N.L., Sen-Mandi, S., and Huang, N. (1997). Selective genotyping, *Mol. Gen. Genet.*, 255: 1–8.

Neeraja, C.N., Maghirang-Rodriguez, R., Pamplona, A., Heuer, A., Collard, B.C.Y., Septiningsih, E.M. (2007). A marker-assisted backcross approach for developing submergence tolerance rice cultivars. *Theo. Appl.Genet.*, 115: 767–776.

Nemoto, K., Ukai, Y., Tang, D.Q., Kasai, Y., and Morita, M. (2004). Inheritance of early elongation ability in floating rice revealed by diallel and QTL analyses, *Theor. Appl. Genet.*, 109: 42–47.

Niroula, R.K., Pucciariello, C., Ho, V.T., Novi, G., Fukao, T., and Perata, P. (2012). Sub1A-dependent and independent mechanisms are involved in the flooding tolerance of wild rice species, *Plant J.*, 72: 282–293.

Okishio, T., Sasayama, D., Hirano, T., Akimoto, M., Itoh, K., and Azuma, T. (2015). Ethylene is not involved in adaptive responses to flooding in the Amazonian wild rice species *Oryza grandiglumis*, *J. Plant Physiol.*, 174: 49–54.

Oladosu, Y., Rafii, M.Y., Abdullah, N. Magaji, U., Miah, G., Hussin, G. et al. (2017). Genotype× environment interaction and stability analyses of yield and yield components of established and mutant rice genotypes tested in multiple locations in Malaysia, *Acta Agric. Scand. Sect. B Soil Plant Sci.*, 67: 590–606.

Pang, Y., Chen, K., Wang, X., Wang, W., Xu, J., Ali, J., et al. (2017). Simultaneous improvement and genetic dissection of salt tolerance of rice (*Oryza sativa* l.) by designed QTL pyramiding, *Front. Plant Sci.*, 8. Doi: 10.3389/fpls.2017.01275

Paterson, A.H., Damon, S., Hewitt, J.D., Zamir, D., Rabinowitch, H.D., Lincoln, S.E. et al. (1991). Mendelian factors underlying quantitative traits in tomato: Comparison across species, generation and environments, *Genetics*, 127: 181–197.

Pradhan, S.K., Chakraborti, M., Chakraborty, K., Behera, L., Meher, J., Subudhi, H.N. et al. (2018).Genetic improvement of rainfed shallow lowland rice for higher yield and climate resilience.https://Krishi.icar.gov.in.

Pucciariello, C., and Perata, P. (2013). Quiescence in rice submergence tolerance: An evolutionary hypothesis, *Trends in Plant Science*, 18: 377-381.

Ramaiah, K., and Ramaswami, K. (1940). Floating habit in rice, *Indian J. Agric. Sci.*, 11: 1–8.

Reynoso, M.A., Kajala, K., Bajic, M., West, D.A., Pauluzzi, G., Yao, A.I., Hatch, K., Zumstein, K., Woodhouse, M., Rodriguez-Medina, J., Sinha, N., Brady, S.M., Deal, R.B., and Bailey-Serres, J. (2019). Evolutionary flexibility in flooding response circuitry in angiosperms, *Science*, 365(6459): 1291–1295.

Sarkar, R.K., Reddy, J.N., Sharma, S.G., and Ismail, A.M. (2006). Physiological bases of submergence in rice and implications for crop improvement. *Curr. Sci.*, 91: 899–906.

Sarkar, R.K., and Panda, D. (2009). Distinction and characterization of submergence tolerant and sensitive rice cultivars, probed by the fluorescence OJIP rise kinetics, *Funct. Plant Biol.*, 36: 222–233.

Sarkar, R.K., and Bhattacharjee, B. (2011). Rice genotypes with Sub1 QTL differ in submergence tolerance, elongation ability during submergence and re-generation growth at re-emergence, *Rice*, **5**(1): 7.

Sasaki, T., Matsumoto, T., Yamamoto, K., Sakata, K., Baba, T., Katayose,Y., et al. (2002). The genome sequence and structure of rice chrosome1. *Nature*, 420: 312–316.

Septiningsih, E.M., Pamplona, A.M., Sanchez, D.L., Neeraja, C.N., Vergara, G.V., Heuer, S. et al. (2009). Development of submergence-tolerant rice cultivars: The Sub1 locus and beyond. *Ann. Bot.*, 103: 151–160.

Septiningsih, E.M., Ignacio, J.C.I., Sendon, P.M.D., Sanchez, D.L., Ismail, A.M., and Mackill, D.J. (2013). QTL mapping and confirmation for tolerance of anaerobic conditions during germination derived from the rice landrace Ma-Zhan red, *Theor Appl Genet.*, 126: 1357–66.

Septiningsih, E.M., Sanchez, D.L. Singh, N., Sendon, P.M., Pamplona, A.M., and Heuer, S. (2012). Identifying novel QTLs for submergence tolerance in rice cultivars IR72 and Madabaru, *Theor. Appl. Genet.*, 124: 867–874.

Septiningsih, E.M., Pamplona, A.M., Sanchez, D.L., Neeraja, C.N., Vergara, G.V., Heuer, S., et al. (2009). Development of submergence-tolerant rice cultivars: the Sub1 locus and beyond. *Ann. Bot.,* 103: 151–160.

Setter, T.L., Ellis, M., Laureles, C.V., Ella, E.S., Senadhira, D., Mishra, S.B. et al. (1997). Physiology and genetics of submergence in rice. *Ann. Bot.*, 79: 67–77.

Siangliw, M., Toojinda, T., Tragoonrung, S., and Vanavichit, A. (2003). Thai jasmine rice carrying QTLch9 (SubQTL) is submergence tolerant, *Ann. Bot.*, 91: 255–261.

Singh, U.S, Dar, M.H., Singh, S., Zaidi, N.W., Barim, M.A., Mackill, D.J. et al. (2013). Field performance, dissemination, impact and tracking of submergence tolerant(Sub1) rice varieties in South Asia. SABRAO J. *Breed. Genet.*, 45: 112–131.

Singh, R., Singh, Y., Xalaxob, S., Verulkarb, S., Yadav, N., Singh, S., et al. (2016). From QTL to variety-harnessing the benefits of QTLs for drought, flood and salttolerance in mega rice varieties of india through a multiinstitutional network. *Plant Sci.*, 242: 278–287.

Singh, N., Dang, T.T.M., Vergara, G.V., Pandey, D.M., Sanchez, D., Neeraja, C.N., et al. (2010). Molecular marker survey and expression analyses of the rice submergence-tolerance gene SUB1A. *Theor. App. Genet.*, 121: 1441–1453.

Singh, S., Mackill, D.J., and Ismail, A.M. (2009). Responses of Sub1 rice introgression lines to submergence in the field: Yield and grain quality, *Field Crops Res.*, 113: 12–23.

Singh, S., Mackill, D.J., and Ismail, A.M. (2011). Tolerance of longer-term partial stagnant fooding is independent of the Sub1 locus in rice, *Field Crop Res.*, 121: 311–323.

Singh, S., Mackill, D., and Ismail, A.B. (2014). Physiological basis of tolerance to complete submergence in rice involves genetic factors in addition to the Sub1 gene, *AoB Plants*, 6: plu 060. https://doi.org/10.1093/aobpla/plu060

Singh, A., Septiningsih, E.M., Balyan, H.S., Singh, N.K., and Rai, V. (2017). Genetics, physiological mechanisms and breeding of flood-tolerant rice (*Oryza sativa* L.), *Plant Cell Physiol.*, 58: 185–197.

Singh, A., Singh, Y., Mahato, A.K., Jayaswal, P.K., Singh, S., Singh, R., et al. (2020). Allelic sequence variation in the *Sub1A, Sub1B* and *Sub1C* genes among diverse rice cultivars and its association with submergence tolerance, *Sci. Rep.*, 10: 8621.

Sripongpangkul, K., Posa, G.B.T., Senadhira, D.W., Brar, D., Huang, N., and Khush, G.S. (2000). Genes/QTL affecting flood tolerance in rice, *Theor. Appl. Genet.*, 101:1074–1081.

Suge, H. (1987). Physiological genetics of internode elongation under submergence in floating rice, *Jpn. J. Genet.*, 62: 69–80.

Suprihatno, B., and Coffman, W.R. (1981). Inheritance of submergence tolerance in rice (*Oryza sativa* L.), *SABRAO Journal*, 13: 98-108.

Sinha, M.M., and Saran, S. (1988). Inheritance of submergence tolerance in lowland rice, *Oryza*, 25: 351-354.

Tang, D.Q., Kasai, Y., Miyamoto, N., Ukai, Y., and Nemoto, K. (2005). Comparison of QTLs for early elongation ability between two floating rice cultivars with a different phylogenetic origin, *Breed. Sci.*, 55: 1–5.

Toledu, A.M.U., Ignacio, J.C.I., Casal, C., Gonzaga, Z.J., Mendioro, M.S., and Septiningsih, E.M. (2015). Development of improved Ciherang-Sub 1 having tolerance to anaerobic germination conditions, *Plant Breed. Biotechnol.*, 37–87.

Toojinda, T., Siangliw, M., Tragoonrung, S., and Vanavichit, A. (2003). Molecular genetics of submergence tolerance in rice: QTL analysis of key traits, *Ann.Bot.*, 91: 243-253.

Tripathi, R.S., and Rao, M.J.B. (1985). Inheritance studies of characters associated with floating habit and their linkage relationship in rice, *Euphytica*, 34: 875–881.

Verma, R., Katara, J., Anilkumar, C., Devanna, B.N., Chidambaranathan, P., Dash, B., et al. (2021). *Advanced Breeding Strategies for Rice Improvement*, 263.

Voesenek, L.A., and Bailey-Serres, J. (2015). Flood adaptive traits and processes: An overview, *New Phytol.*, 206: 57–73.

Wang, Y. (2009). Screening of germplasm adaptable to direct seeding and discovery of favorable alleles for seed vigor and seedling anoxic tolerance in rice (*Oryza sativa* L.), Nanjing: Dissertation; Nanjing Agricultural University; 2009.

Wang, N., Ding, L.J., Xu, H.J., Li, H.B., Su, J.Q., and Zim, Y.G. (2015). Variability in responses of bacterial communities and nitrogen oxide emission to urea fertilization among flooded paddy soils. *Fems.Microbiol.Ecol.*, 91: 11–18.

Xu, K., Xu, X., Fukao, T., Canlas, P., Maghirang-Rodriguez, R., Heuer, S. et al. (2006). Sub1A is an ethylene-response-factor-like gene that confers submergence tolerance to rice, *Nature*, 442: 705.

Xu, K., Xu, X., Ronald, P.C., and Mackill, D.J. (2000). A high–resolution linkage map of the vicinity of the rice submergence tolerance locus Sub1. *Mol. Gen. Genet.*, 263: 681–689.

Xu, K., and Mackill, D.J. (1996). A major locus for submergence tolerance mapped on rice chromosome 9. *Mol. Breed.*, 2: 219–224.

Yamauchi, M., Aguilar, A.M., Vaughan, D.A., and Seshu, D.V. (1993). Rice (*Oryza sativa* L.) germplasm suitable for direct sowing under flooded soil surface, *Euphytica*, 67: 177–84.

Young, N.D. (1996). QTL mapping and quantitative disease resistance in plants, *Annu. Rev. Phytopathol.*, 34: 479–501.

Zhang, M., Lu Q., Wu, W., Niu, X., Wang, C., Feng, Y., et al. (2017). Association mapping reveals novel genetic loci contributing to flooding tolerance during germination in indica rice,*Front. Plant Sci.*, 8: 678.

Zhao, Q., Feng, Q., Lu, H., Li, Y., Wang, A., Tian, Q., et al. (2018). Publisher Correction: Pan-genome analysis highlights the extent of genomic variation in cultivated and wild rice, *Nat. Genet.*, 50: 1196.

Zhao, Q., Feng, Q., Lu, H., Li, Y., Wang, A., Tian, Q. et al. (2018). Pan-genome analysis highlights the extent of genomic variation in cultivated and wild rice, *Nat. Genet.*, 50(2): 278–284.

Zhou, L., Lu, Y., Zhang, Y., Zhang, C., Zhao, L., Yao, S., et al. (2020). Characteristics of grain quality and starch fine structure of japonica rice kernels following pre-harvest sprouting, *J. Cereal Sci.*, 95: 103023. doi: 10.1016/ j.jcs.2020.103023 Z.

7

Yield and Yield-Gap

Rice is a semi-aquatic crop; however, the ability to survive complete or partial flooding is limited for most of the rice varieties. On the other hand, there are traditional varieties that have adapted well in flooded environments. These varieties are low yielding and have poor grain quality in some cases. The International Rice Research Institute (IRRI) has been studying the genetics underlying the trait of these traditional rice varieties that can tolerate higher level and longer duration of flooding.

Stagnant flooding is a condition where water stays more than a month and the lines remain partially submerged. Complete submergence usually happens at seeding-vegetative stage due to typhoons or heavy rains. The anaerobic germination happens in a direct seeding method of planting seed and the field gets flooded, resulting to poor germination of the rice seed.

This introduced the developmental genetics and its molecular ecological basis of high-yielding forms of rice in the past decade, and analyzed the advantage and the shortage of comparative physiological approach traditionally used in research work on crop cultivation. Need was emphasized for actively introducing the research content and its methodology from relative disciplines to deeply understand the scientific issue, and suggested that the key to realize stable and high-yielding rice was to develop a rational cultivation system based on the properties of genetic effects on the traits at different developmental stages by controlling and regulating the traits governed by dominant-effect genes and additive-effect genes x environment in same direction, which was considered as the main characteristic and technological innovation of modern crop genetic and ecological cultivation science. Finally, the development trend of crop cultivation science shifting to molecular crop cultivation science was predicted and discussed.

Analysis on panicle dry weight at different developmental stages based on two years of experimental data was conducted by using additive-dominant developmental genetic models with genotype and genotype × environment interaction effects. The results showed that panicle dry weight was controlled by both genetic main effects and genotype × environment interaction effects in the whole developmental process. During the early-middle developmental periods (15 days after flowering), panicle dry weight was mainly governed by dominant effects, and dominant genes were expressed with large amounts. Meanwhile, positive effects produced by environmental factors appeared to be significant. Proper cultivation measures to create good external environment conditions for rice panicle was helpful to promote the expression of heterosis potential completely. Within the middle-late developmental periods (from the fifteenth day to the thirtieth day after flowering), additive effects played a major role, and its genes were expressed in the most active state. Genetic selection for panicle conducted in this period was expected to lead to better genetic advancement.

The models of genetic effects and genotype × environment interaction for additive-dominant-additive x additive epistasis were used to analyze the panicle traits of inter-sub-specific crosses of rice (*Oryza sativa*) in different environments. It was found that significant additive, dominant and additive x additive epistatic effects and genotype x environment (GE) interaction was observed in main panicle length, spikelet density, primary branches, total lengths of primary and secondary branches, but the numbers of main panicle and secondary branches showed no significant additive × environment (AE) interaction and dominant × environment (DE) interaction. The seven traits studied were mainly controlled by dominant effects, but branches' traits were more obviously affected by DE interaction. Heritability analysis showed that general heritability in a broad sense (HG2) was much larger than other heritabilities. To a certain extent, the interaction heritabilities showed their effects in seven panicle traits that were tested. Heterotic prediction indicated a positive heterosis in all panicle traits except the numbers of primary and secondary branches. GE only influenced the expression extents of heterosis but was not able to change their direction. According to predicated genetic effects, IR66158-37, IR65600-85, Minhui63 and R669 were better than other parents in the tactics of breeding for improving panicle traits because the progenies from these crosses always showed that panicle traits were slightly affected by the environment (Sharma and Jaiswal, 2021).

The average yield potential of these landraces is 2 t/ha as compared to 6–8 t/ha of semi-dwarf commercial varieties. Unfortunately, the 'semi-dwarf varieties', which are widely cultivated in Asia, are susceptible to flooding and die in a few days after submergence. The submergence-tolerant varieties have resulted in a considerable increase in rice production and income to farmers in flood-prone areas (Ismail et al., 2013). In India, rice varieties, such as Sabita, Ambika, Saraswati, and Hangseswari are recommended by the Indian Council of Agricultural Research for cultivating deep-water rice. These varieties, such as Hangseswari, can produce up to 2.5 t/ha grain yield under deep-waterlogged conditions (www.icar.org.in).

7.1 Flash Flood

The main strategies enabling rice plants to cope with flash flooding stress require growth regulation during submergence and subsequent rapid growth recovery after de-submergence. Further, better nutrient management options can enhance this strategy. Application methods of nitrogen and phosphorus were evaluated in submerged rice for tiller mortality, productivity, grain quality, and nutrient absorption. The performance of Sub1 (IR-64 Sub1 and Swarna Sub1) and non-Sub1 (IR-20) cultivars of rice was tested under clear and turbid water for their tolerance to submergence in response to basal phosphorus and post-submergence nitrogen. Tillering ability, yield, and nutrient absorption of rice subjected to complete submergence for 15 days decreased significantly over non-submerged rice plants. Turbid water submergence was fatal in terms of tiller mortality, reduced nutrient absorption, and yield because of low light and dissolved oxygen underwater. The crop fertilized with nitrogen produced greater number of tillers, yield attributes, and grain yield than the unfertilized crop under complete submergence at maximum tillering stage. The effects were more positive when basal phosphorus was applied with nitrogen. Urea foliar spray after de-submergence significantly enhanced the leaf area, dry matter weight, specific leaf weight, tiller regeneration ability, and narrowed down the flowering time which led to higher grain yield and productivity. The findings of the study suggest that higher rice yield can be obtained by adjustment in timing of nutrient application in flood-prone areas, where farmers are scared to apply fertilizer because of the risk involved.

Flash floods adversely affect rice productivity in vast areas of rain-fed lowlands in South and Southeast Asia and tropical Africa. Tolerant landraces that withstand submergence for one to two weeks were identified; however, incorporation of tolerance in modern high-yielding varieties through conventional breeding methods has been slow because of the complexity of both the tolerance phenotype and flood-water conditions, and the ensuing discrepancies encountered upon phenotyping in different environments. Designing an effective phenotyping strategy requires a thorough understanding of the specific flood-water characteristics that most likely affect survival during flooding. Researchers have investigated the implications of flood-water temperature and light penetration, caused by artificial shading, seasonal variation, or water turbidity, for seedling survival after submergence. Three field experiments were conducted using rice genotypes contrasting in their tolerance of submergence: FR13A and Kusuma (tolerant); Gangasiuli (intermediate); Sabita, CRK-2-6 and Raghukunwar (elongating/avoiding types); and IR42 (sensitive). When tested, the hypotheses was that warmer flood-water decreases plant survival and that turbid water augment plant mortality by causing effects similar to those caused by shading, by reducing light penetration. Plants survive better when water is cooler, and survival decreased by about 8% per unit increase in water temperature above 26°C. Lower intensity of light and warmer temperatures seem to reduce biomass and increase mortality under flooding. An increase in the concentrations of O_2 and CO_2 and a decrease in water pH did not improve survival in clear unshaded water. Turbid flood-water was more damaging to rice as plant mortality increased with the percentage of silt increase, and the effects of water turbidity cannot be explained by reduction of light penetration alone. Even the most tolerant rice cultivar, FR13A, experienced higher mortality when flooded with turbid flood-water. Correlation studies revealed that cultivars with the capacity to maintain higher biomass, higher chlorophyll, and non-structural carbohydrate concentrations after submergence had higher survival. These findings help to understand the variations observed in submergence tolerance when screening is done under different environments. The study could have implications for designing proper screening strategies and assessing the damage submergence causes across different rice-growing regions (Das et al., 2009).

Rice plants may experience submergence stress in various phases of plant development due to unpredictable flash floods occurring during planting season. Plants' recovery phase after submergence is an important phase that determines plant survival. The character of plant growth and development during this recovery phase can be important information in determining the tolerance of plants to submerged stress. Therefore, this study was conducted to evaluate the recovery phase of post-submerged rice plants indicated by the character of plant growth and development. Split plot design was utilized with rice cultivars as the main plot and submergence time subplot. Two cultivars used were IR 42 (non-tolerant cultivar) and Inpari 30 (tolerant cultivar), while submergence time consisted of three phases, namely: (1) generative initial phase = 42 days after transplanting (DAT); (2) active tillering phase = 28 DAT, and (3) early vegetative phase = 7 DAT. Plant samplings were conducted six times: right before submergence (t0), right after submergence (t1), four days after submergence (t2), seven days after submergence (t3), 10 days after submergence (t4) and 14 days after submergence (t5). Results showed that both cultivar and submergence time showed a non-significant effect on plant growth during recovery phase, except for leaf greenness parameter.

The recovery pattern of rice plants after experiencing submergence stress was generally not influenced by the age of the plants during stress. In addition, improvements in plant growth characters began to appear after 10 days of recovery (Irmawati et al., 2020).

Early season flash flooding and submergence greatly impair rice production in the rain-fed lowlands of West Africa. Here, rice cultivation is through transplanting in puddled soil. Crops established during the early part of the rainy season are adversely affected by submergence, while delayed transplanting until the flood waters have receded results in the use of old seedlings and terminal drought stress. While the use of submergence tolerant (Sub1) rice varieties can greatly reduce the yield loss caused by one to two weeks of early season submergence, changing to wet seeding at the start of the rainy season may confer additional advantages. Elsewhere, wet seeding has been shown to enable timely (earlier) crop establishment and more rapid early growth, and thus the potential to provide greater resilience to submergence. Therefore, the objective of this study was to assess the performance of two Sub1 varieties developed for West Africa (FARO 66, FARO 67) under transplanted and wet-seeded conditions, in comparison with the predominant local variety (WITA 9). On-station experiments were carried out over three seasons (dry 2018, wet 2019, dry 2019) at Mbé Research Station of the Africa Rice Center, Bouake, Cote d'Ivoire. The fields were submerged for one to two weeks at five to seven weeks after seeding, and the wet season crop was sown at the beginning of the rainy season. Yield of the Sub1 varieties was 1.1–4.5 t ha^{-1} higher than that of WITA 9, depending on the season and establishment method. Wet seeding resulted in much higher yields of WITA 9 than transplanting in the wet season but yield of the Sub1 varieties was not affected by establishment method in any season. However, tiller survival and biomass production of the wet seeded Sub1 crops were less affected by submergence than the transplanted crops. Wet seeding also reduced labour requirement and cost, and increased profitability. Furthermore, establishment of the wet season crop at the beginning of the rainy season facilitated intensification to two crops per year. Therefore, the adoption of Sub1 varieties could enable significant progress towards the goal of self-sufficiency in rice production in West Africa. Combining Sub1 with wet seeding would provide further benefits in terms of increased profitability and resilience to flash flooding (Devkota et al., 2022).

7.1.1 Rationale of Yield Gap

Since 1990, rice production has grown at a lower rate than the population. This deceleration in the growth of rice production is a cause for concern in terms of world food security. It has been the topic of numerous reviews and several rice scientists have warned of the risk of a pending food crisis. Based upon the 1997 estimated production of 524 million tonnes, an additional 180 million tonnes of paddy rice will be required in the year 2025. Most of the increased demand will be required by developing countries.

Yield gaps are still observed in many countries, while evidence of productivity decline in intensive rice production has been increasingly noticed both at research stations and farmers' fields. Due to its complexity, there are different points of view regarding the possibility of narrowing the yield gaps as a tool for increasing rice production. A number of experts are of the opinion that yield gaps in favorable rice ecologies are not significant in exploitation for increasing rice yield and production. Others believe that large rice yield gaps still exist. In 1995, the average rice yields in 78 countries were less than the world average yield of 3.77 t/ha, indicating the existence of yield gaps. Also, progressive farmers usually obtain higher yields and more profits than the ordinary farmers.

There is little doubt that intensified rice cropping has significant effects on soil chemical and biological processes. The chemistry of submerged soils is complex and the addition of organic matter from crop residues and continuous soil reduction from puddling further complicates the soil chemical and biological processes. Moreover, the socio-economic factors and the degradation of irrigation infrastructures may contribute to yield gaps, thereby bringing about productivity decline under farmers' field conditions.

Narrowing the yield gaps aims not only to increase rice yield and production but also to improve the efficiency of land and labour use; also, to reduce the cost of production and increase sustainability. Exploitable yield gaps of rice are often caused by various factors which may be classified into physical, biological, agronomic, socio-economic and institutional constraints. These can be effectively improved through participatory approach in action. The narrowing of the yield gap is not static but dynamic with the technological developments in rice production, as the gaps tend to enlarge with increase in yield potential due to the use of improved varieties.

The narrowing of the yield gap of rice requires integrated and holistic approaches, including appropriate concept, policy intervention, understanding of farmers' actual constraints in achieving high yield, deployment of new technologies and promotion of integrated crop management, adequate supplies of inputs and farm credit, and strengthening of research and extension and the linkages between them. If one of these components is missing or is weak, the narrowing of the yield gap in a particular rice production area cannot reach its full potential.

The causes of yield gaps in rice differ widely from season to season, country to country and/or even from location to location within a country or region. It is essential, therefore, to promote close collaboration between research, extension, local authorities, non-governmental organizations (NGOs) and the private sector in order to identify specific constraints in high yield and appropriate technologies and solutions and take concerted action to bridge the yield gaps in rice through participatory approaches. This will depend mainly on the will of the governments to support, co-ordinate, and monitor such integrated and holistic programmes. International support to governments' initiatives in this direction could speed up sustainable increased rice production and the conservation of natural resources and the environment for the benefit of future generations (FAO, 1999).

7.1.2 Long-term Submergence/Deep-water Rice

Submergence is an important constraint which prolongs partial-submergence damage of rice plants and reduces grain yield. Due to the heterogeneity in flood-prone ecosystem, many different types of traditional rice varieties are being grown by the farmers. The local landraces adapted to extremes in water availability could be the sources of new gene(s) which would be utilized to improve the adaptability of rice to submergence with high yield. The main goal of this study is to identify new genetic resources which are tolerant to submergence based on morpho-physiological traits. Field experiments comprising 22 varieties, which were completely submerged for 15 days under RBD design, were conducted to screen out the submergence-tolerant varieties based upon morpho-physiological characters in the costal ecosystem of Odisha during *kharif* 2014 at the Adaptive Research Station, Sakhigopal. Nitrogen content of before submergence and during harvesting were the key criteria of yield. Nitrogen uptake during harvesting is directly proportional to the yield. There are different yield-attributing characters, like 1000 grain weight, number of panicles per unit land and number of fertile grains/panicles which influence the final yield of the variety. The suitable combination among these varies depends upon the submergence tolerance of the variety. Based on the findings, it was concluded that the grain yield is positively correlated with the number of grains/panicle and filled grains/panicle and negatively correlated with sterility percentage, which is found in case of Sabita, FR-43B and Jalamagna (Mahapatra, 2017) (Table 7.1).

The effect of waterlogging on yield is summarized in Table 7.2. The yields in all treatment groups significantly decreased as compared with that of the control group, except for the 1-d 1/1 submergence group, which showed a significant increase in yield (7.9%). Wang et al. investigated the influence of slight submergence (two and four days) on mid-season rice at the final phase of the tillering stage and found that the yields of the experimental groups were close to those of the control group (Wang et al., 2014).

Table 7.1: Yield and its attributing characters in response to submerged condition of rice varieties.

Variety	Panicle Length (cm)	Grains/ Panicle	Filled Grains /Panicle	Sterility (%)	1000 Grain wt. (g)	HI (%)	Yield (q/ha)
Sabita	26.68	156.0	86.80	44.30	25.70	32.54	38.43
FR43B	24.08	162.00	83.00	48.70	24.70	32.46	38.07
Jalamgna	20.56	155.37	80.00	48.50	23.54	31.34	36.72
OR2331-14	22.91	209.53	99.80	52.30	20.27	30.76	36.60
IR58058-Sub 17	26.80	179.90	80.00	55.5	20.60	30.37	36.63
Jayanti Dhana	23.61	180.77	78.40	56.6	27.67	29.87	35.40
Jalamani	22.40	154.50	68.00	55.9	26.35	29.55	35.2
CRDhan-500	23.30	159.80	72.50	54.6	26.34	29.36	34.54
CRDhan-401	23.94	164.67	66.80	59.4	24.85	29.20	35.23
CRDhan-505	24.03	161.80	68.70	57.5	25.58	28.89	34.90
Mahalaxmi	24.50	198.47	88.00	55.3	23.57	28.94	34.48
Monika	22.25	191.47	81.0	57.6	20.74	28.39	33.48
CRDhan1030	25.36	175.27	71.30	59.3	20.50	28.22	33.53
OR-142/99	21.49	179.60	69.20	61.4	20.54	27.90	33.59
Tanmayee	24.20	148.13	62.60	57.7	25.24	27.84	33.57
Urbashi	20.39	165.90	62.00	62.6	20.61	27.53	32.17
Salibahan	23.47	172.10	63.60	63.0	23.33	26.92	31.42
Rambha	25.42	156.87	59.40	62.1	24.50	26.33	31.00
OR/2328/05	21.60	165.47	59.00	64.3	22.18	25.16	29.47
Mayurakantha	22.57	200.17	72.00	64.0	22.64	24.80	27.13
Kalasira	24.75	146.50	55.00	62.4	27.26	24.31	26.00
Bankoi	22.13	152.07	52.80	65.0	24.73	23.94	26.10
CD5%P	1.41	3.12	2.91	3.01	2.36	0.29	1.35
CV	3.63	1.11	2.60	3.15	6.05	0.61	2.45

Li et al. only sporadically observed yield increases after waterlogging: After neck submergence, rice shows only minor underproduction, and either no underproduction or a production increase is observed, at the stage of rice during waterlogging, and the cultivar is tested. The temperature of the flood-water is close to the optimal temperature range of rice (Li et al., 2004). The differences between this results and the results reported in the literature (Wang et al., 2014; Li et al., 2004) may be due to differences in the rice waterlogging duration, with the waterlogging temperature findings indicating that short-term (1 day) complete submergence significantly

Table 7.2: Effects of submergence on rice yield, yield components, and effective leaf area in plants at the tillering stage.

Submergence depth-duration (d)	Panicle number per pot	Grain number per panicle	Seed-setting rate (%)	1000-seed weight (g)	Yield (g/pot)	High-tilling panicles per pot	Photosynthetic rate (μmol·cm^{-2}·s^{-1})	Photosynthetically effective leaf area (cm^2/ stem)
CK	27.1 bc	105.6 a	91.1 bc	32.5 a	84.6 b	0.0 d	27.9 ab	85.1 b
1/2-1	26.8 bc	99.1 b	93.4 a	32.1 a	79.5 c	0.0 d	27.2 b	77.0 c
1/2-3	27.5 bc	98.4 b	88.9 de	32.0 a	77.0 d	0.3 cd	26.5 bc	74.7 cd
1/2-5	27.3 bc	99.2 b	86.3 f	32.2 a	75.2 d	1.3 bc	26.1 bc	73.7 cd
2/3-1	26.3 c	98.0 b	92.8 ab	32.1 a	76.8 d	0.3 cd	27.7 ab	87.8 ab
2/3-3	27.2 bc	94.3 c	90.3 cd	32.5 a	75.3 d	0.5 cd	25.4 c	74.6 cd
2/3-5	27.8 bc	93.2 c	88.1 ef	31.7 a	72.3 e	1.7 b	25.0 c	73.3 cd
1/1-1	28.2 b	105.8 a	93.4 a	32.8 a	91.3 a	0.5 cd	28.6 a	89.6 a
1/1-3	31.8 a	88.0 d	88.0 ef	32.1 a	79.1 c	1.2 bc	24.8 c	72.7 cd
1/1-5	32.0 a	87.5 d	87.7 ef	31.3 a	76.9 d	2.8 a	23.4 d	71.4 d

Values followed by a different letter are significantly different (*t*-test, $P \leq 0.05$).

doi:10.1371/journal.pone.0127982.t001

increases rice yields. This result was presumably associated with a compensatory or super-compensatory effect in the plant upon removal of the waterlogging stress. Super-compensatory effects are a universal phenomenon in biology and are normally caused by stress and damage as a self-accommodation behavior in response to harmful environments. For some crops, the compensatory effects caused by moderate stress are even better than the effects of normal growth conditions, resulting in increased production or minor underproduction (Shan et al., 2000). Furthermore, water deficit does not always lead to crop underproduction; indeed, deficit in early stages may lead to production increase in some crops (Turner, 1997) (Table 7.2).

7.2 Combined Effects of Submergence and Nutrition

One experiment was undertaken with the treatments including T1: 0.0 mg ZnSO4 per kg soil, T 2: 5.0 mg ZnSO4 per kg soil, T3: 10 mg ZnSO4 per kg soil. Treatments were done in submerged and non-submerged (controlled) pots, both at the time of sowing. After 30 days of sowing, 15 days of complete submergence was maintained. This experiment was conducted to study the effect of zinc sulphate by assessing yield attributes, at before and after submerged condition, for improving submergence tolerance in Sub1 and non-Sub1 rice varieties. A completely randomized block design was adopted with three treatments, three replications and five rice varieties selected on the basis of submergence tolerance and susceptibility. Seeds were sown in pots directly. It was concluded that zinc sulphate @10 mg/kg soil significantly enhanced the yield of Sub1 and non-Sub1 rice varieties even after 15 days of complete submergence (Singh et al., 2020) (Table 7.3).

A field experiment was carried out during *kharif*, 2016 at the farm of ICAR-Krishi Vigyan Kendra (KVK), Mangaluru, Karnataka state to study the effect of Neem cake-urea mixed application on growth and yield of rice under submerged condition. The soil of the experimental field was lateritic with acidic pH, medium in available nitrogen, high in phosphorus and low in potassium content. The experiment was laid out in randomized block design with seven treatments and each one was replicated thrice. The treatments composed of absolute control, 100% nitrogen along with five treatments of urea and Neem cake mixed in different proportions. The results indicated that number of the tillers per m^2, number of panicles per m^2 and plant height (cm) were significantly higher in treatment T3 comprising 100% RDN in three splits with 100% Neem cake application followed by treatment with T4 comprising 75% RDN with 100% Neem cake application compared to other treatments. The grain yield and straw yield were significantly higher in T3-RDN (100%) in three splits + Neem cake (100%) followed by T4-RDN (75%) + Neem cake (100%). The lowest

Table 7.3: Effect of zinc sulphate on grain yield per plant (g) and test weight (g) on Sub1 and non-Sub1 rice varieties exposed to 15 days of complete submergence.

Variety	Grain Yield/Plant (g)	Test Weight (g)
	T1 T2 T3	T1 T2 T3
	A B A B A B	A B A B A B
Swarna	20.8317.83 21.7918.3922.7518.92	19.7518.6019.6818.9419.6819.63
Swarna Sub1	22.9216.3323.6516.5524.5017.33	19.8519.4620.6620.1620.9820.37
Samba Mahsuri	24.8518.1625.1019.0825.8420.59	19.3918.5120.6218.8021.0319.71
Samba Mahsuri Sub1	26.7922.8728.1223.3028.8424.09	20.5618.9021.5118.9521.7319.91
IR64sub1	17.9321.6818.3122.7019.5923.35	19.5818.5119.7418.8220.1519.11

A = Controlled condition B = Submerged condition

Table 7.4: Effect of Neem cake-urea mixed application on growth and yield of rice under submerged condition.

Treatment	Plant Ht. (cm)	Tillers per m^2	Panicles per m^2	Grain Yield q/ha	Straw Yield q/ha
Control (N0P30K60)	93.0	320	286	38.0	3.46
RDF, N60P30K60	100.3	421	380	42.0	4.07
100 % RDN + Neem cake (100%)	107.3	472	445	55.0	4.35
75 % RDN + Neem cake (100 %)	106.4	468	442	52.0	4.28
50 % RDN + Neem cake (100 %)	103.0	428	410	46.0	4.13
50 % RDN + Neem cake (75 %)	102.0	426	398	45.0	4.12
50 % RDN + Neem cake (50 %)	102.0	419	376	43.0	4.10
S. Em.	0.63	11.1	9.44	1.31	0.04
CD (5 %)	1.85	33.0	28.0	3.88	0.13

grain yield and straw yield was recorded in treatment T1-control where nitrogen was not applied (Shenoy, 2020).

Intensive tillage practices along with improper residue management in a rice (*Oryza sativa*) system (RRS) contributed to soil fatigue and declining productivity in South Asia. Therefore, a three-year (2013–2015) field study was conducted to assess tillage modification effects on productivity and soil C sequestration under RRS at ICAR - RC for northeastern hill region, Tripura, India. The experimental site represented two different ecologies: unsubmerged (ECO 1) and submerged (ECO 2), with three tillage practices: conventional tillage (CT), reduced tillage (RT), and no-till (NT). Results showed that the cultivation of RRS under RT produced significantly higher grain (8.6–9.4 Mg ha^{-1}) and straw (11.8–12.9 Mg ha^{-1}) yields under both ecologies over those under CT and RT, in addition to recycling the maximum biomass. Soil under RT had lower bulk density (ϱ_b), the highest soil organic carbon (SOC) concentration, pool, sequestration, accumulation, carbon retention efficiency, soil microbial biomass carbon, and dehydrogenase activities under both ecologies as compared to CT. A total amount of 1.34 Mg C ha^{-1} was accumulated under soils of RT over three years. The rate of SOC sequestration ranged from 133.6 kg ha^{-1} year^{-1} under soils of CT to 444.7 kg ha^{-1} year^{-1} under RT in RRS. Thus, cultivation of RRS under RT with effective residue recycling is recommended for higher system productivity and C sequestration under both rice production ecologies of the NE hilly region of India (Yadav et al., 2019) (Table 7.5).

This study explored the response in grain yield and quality of upland and wetland rice varieties to a combination of zinc (Zn) and nitrogen (N) fertilizers under two water-management regimes. A factorial arrangement based on a randomized complete block design composed of three factors was carried out with three independent replications. Upland and wetland rice varieties were grown with three fertilizer treatments: the optimum N rate (86 kg N ha^{-1}) without Zn application, the optimum N rate with Zn (50 kg $ZnSO_4$ ha^{-1}), and the high N rate (172 kg N ha^{-1}) with Zn

Table 7.5: Grain and straw yield of rice grown under no tillage and different eco-systems.

Treatment	DR grain yield (Mg/ha)								
	2013			2014			2015		
	ECO 1	ECO 2	Mean	ECO 1	ECO 2	Mean	ECO 1	ECO 2	Mean
CT	4.80	3.50	4.15	5.07	3.70	4.38	5.10	3.90	4.50
RT	5.10	3.67	4.38	4.90	3.80	4.35	5.33	4.10	4.72
NT	4.10	3.53	3.82	4.00	3.60	3.80	4.00	3.47	3.73
Mean	4.67	3.57		4.66	3.70		4.81	3.82	
	ECO	Till	ECO × Till	ECO	Till	ECO × Till	ECO	Till	ECO × Till
$LSD_{0.05}$	0.16	0.20	NS	0.22	0.27	NS	0.27	0.33	NS
	DR straw yield (Mg/ha)								
	2013			2014			2015		
	ECO 1	ECO 2	Mean	ECO 1	ECO 2	Mean	ECO 1	ECO 2	Mean
CT	6.50	4.77	5.63	6.83	5.03	5.93	6.90	5.30	6.10
RT	7.00	4.93	5.97	7.10	5.13	6.12	7.20	5.53	6.37
NT	5.40	4.80	5.10	4.47	4.90	4.68	5.40	4.67	5.03
Mean	6.30	4.83		6.13	5.02		6.50	5.17	
	ECO	Till	ECO × Till	ECO	Till	ECO × Till	ECO	Till	ECO × Till
$LSD_{0.05}$	0.35	0.43	NS	0.09	0.11	NS	0.29	0.35	NS

No-Till; ECO: Ecology

under waterlogged and well-drained conditions. Grain yield was 27% lower in the well-drained than in the waterlogged condition in wetland rice, while there was no effect in upland rice. Application of optimum N with Zn produced the highest grain yield in upland rice, while yield was the highest in wetland rice in high N with Zn application. Upland rice grown in well-drained condition with optimum and high N with Zn treatments enhanced Zn concentration by 45% and 29% higher than the treatment without Zn, respectively, while it had no difference among three treatments in the waterlogged condition. Wetland rice variety grown under well-drained condition in optimum and high N rate with Zn treatments was equally effective in improving grain Zn concentration at the average of 88% as compared to the control; while rice grown under the waterlogged condition in high N with Zn treatment improved 92% the concentration. The optimum N rate with Zn application increased grain yield in upland rice, while higher N input is required for wetland rice. Grain Zn concentrations of upland and wetland rice varieties were enhanced by applying Zn fertilizer; however, the increased level depended on N application rate in an individual water condition (Yamungmorn et al., 2020).

Cold waterlogged paddies typically have low-to-average yield due to their relatively low soil temperature, poorly developed plough layer, and lack of available nutrients. Despite the above, yield can be improved by targeted measures, such as supplementing with special organic materials. Past research has shown that biochar can play an important role in sequestering soil C, building soil fertility, and improving crop yield. However, the effects of biochar on crop yield and soil properties in cold waterlogged paddies have not been thoroughly investigated. It was hypothesized that biochar improves soil fertility, and thus, increases rice grain yield in cold waterlogged paddies. A two-year field experiment was conducted in 2011 and 2012 to investigate the effect of bamboo biochar (BB), rice straw biochar (RB), and rice straw (RS) on soil's physical and chemical properties, grain yield, and yield components in a cold waterlogged paddy in Zhejiang province, China. Results showed that both BB and RB significantly increased soil pH and soil organic carbon compared to control, whereas their effects on total N were either very small or insignificant. Application of RB significantly increased soil P and K in both years, and the increases relative to control were greater in 2011 (by 33.9% and 99.1%, respectively) than in 2012 (by 15.3% and 28.6%, respectively). Moreover, RB application resulted in greatest improvement in grain yield (8.5–10.7% greater than that from the control), and this may be attributed to increased nutrient availability (mainly P and K). Yield component analysis indicated that experimental treatments had the greatest effect on thousand-grain weight, followed by the number of productive tillers per plant and harvest index. Neither biochar nor RS significantly affected the total nutrient (N + P_2O_5 + K_2O) content of grains, although the K content of grains from BB and RB plots was significantly higher than in those from control plots in 2012. The total nutrient content of straw under RB treatment was significantly higher than that under control and RS treatments in 2012, mainly due to increased K content (by 12.0%) of straw. The total nutrient uptake by grain was significantly (13.6–16.4%) higher under RB treatment than under control treatment. This was primarily due to the relatively high K uptake by RB grains (15.9–22.6% greater than that by the controls). Similarly, the total nutrient uptake by straw from RB plots was significantly greater than that of straw from the control plot. Further studies on biochar in cold waterlogged paddies are essential in order to evaluate the long-term effects of biochar and its behavior in soils (Liu et al., 2016).

7.3 Sub1 Varieties and Mega Varieties

Experiments were conducted in 2014 and 2015 *kharif* season to evaluate the performance of rice genotypes under submerged condition on the basis of morpho-physiological traits and yield attributes. The study revealed that rice var. Swarna Sub 1 exhibited the highest survival (91.2%), whereas var. IR42 as susceptible check showed the least survival (4.2%) under submergence. Among the tested genotypes, IR 10F365, IR 11F216, and IR 11F239 with the respective survival values of 81.4, 80.0, and 78.1%, respectively, were found to be at par with the var. Swarna Sub1 after 16 days of complete submergence at vegetative stage. Moreover, physiological traits, like chlorophyll, sugar content, and anti-oxidative system (SOD and CAT) were higher in tolerant genotypes as compared to susceptible ones. Less reduction in the content of sugar, chlorophyll, SOD, and CAT activity was observed in genotypes IR 10F365, IR 11F216, and IR 11F239 along with var. Swarna Sub1 (tolerant check) after submergence, while in susceptible genotypes IR 09L311, IR 08L216, and IR 55423-01, the reduction in sugar and chlorophyll content was higher just after submergence. Apart from physiological traits, tolerance genotypes have higher yield and yield-attributing character as compared to susceptible genotypes. Thus, the study revealed that not only var. Swarna Sub1 even few more elite rice lines were found superior in terms of submergence tolerance and yield (Dwivedi et al., 2017).

Rice genotypes IR 10 F365 (24.5 g plant^{-1}), IR 10L182 (39.3 g plant^{-1}), IR 11F216 (27.1 g plant^{-1}) and IR 11F239 (21.1 g plant^{-1}) have higher grain yield as compared to tolerant check cultivar var. Swarna Sub1 (10.10 g plant^{-1}) under short-term flooding condition, suggesting that genetic variation in tolerance of transient submergence is present in rice genotypes.

Table 7.6: Performance of genotypes after submergence stress for 16 days at vegetative stage.

Genotype	Days to 50%Flg.	Pl.ht. (cm)	Panicles/ plant	Panicle length (cm)	Fert. Grains/ panicle	Fertility %	1000-grain wt.(g)	Grain yield (g/ plant)
IR 07L320	126	123	14.0	26.1	136	82.1	23.6	30.5
08L216	126	100	10.6	26.1	159	89.3	22.3	27.1
09L204	130	93	15.9	23.9	118	74.3	21.1	29.3
09L261	132	91	16.7	23.3	148	81.4	17.1	33.4
09L311	131	104	14.0	24.5	119	73.6	28.3	30.3
09L337	128	105	15.6	25.1	159	78.1	23.4	35.7
09L342	129	109	13.9	25.1	170	89.5	22.3	33.9
10F365	128	100	11.1	27.0	135	84.1	23.6	24.5
10F602	130	93	13.0	26.5	119	65.4	21.9	20.6
10L182	126	104	13.7	24.4	154	85.8	23.3	39.3
11F190	131	108	10.1	22.2	115	72.3	19.4	14.9
11F195	141	109	9.4	21.9	159	69.1	19.2	17.9
11F216	141	103	9.9	22.3	76	48.4	17.5	11.7
11F239	140	92	13.5	25.3	124	76.5	23.5	21.0
IR55423-01	127	108	13.2	27.1	103	62.3	22.6	25.6
Swarna	144	79	10.2	18.4	53	50.2	16.1	8.77
Swana Sub1	140	81	12.2	21.5	86	65.7	18.5	10.1
Mean	132.7	100.2	12.8	24.2	125.4	73.6	21.4	24.4
C.D.(05)	4.49	6.16	4.57	2.43	32.56	13.2	2.94	7.09

Conventional rice Xiangzaoxian 45 was used as test material and waterlogged at different submergence depths (two-thirds submerged and fully submerged) for variable durations (three, five, seven and nine days) at the jointing stage to investigate the influence of waterlogging stress on green leaf number, curly leaf' length, leaf sheath length, high tillering and other morphological factors and yield components. Results showed that, in the different submergence depth treatments, green leaf's number had a negative correlation with the flooding time, and two-thirds of submergence had a more obvious effect than full submergence. Stems changed obviously in full-submergence treatment, while leaves changed significantly in two-thirds of submergence. High tillering rate had a positive correlation with the flooding time. Curl leaf's length had a negative correlation with the flooding time under two-third submergence, but a positive correlation under full submergence. Leaf sheaths among different treatments, all showing elongation growth, were not significantly different. The different submergence stresses increased the unfulfilled grain rate and empty grain rate and lowered the thousand-grain weight. Treatment of two-thirds of submergence for nine days had the severest effect, with lowest grain number (34). It was discussed the adaptation mechanisms and possible causes for stems, leaves, and other morphological characteristics under different submergence stresses. The results could provide a scientific basis for flooding disaster reduction and disaster-resistant breeding of rice for the middle and lower reaches of Yangtze river (Jin-hu, 2014).

Super rice, Lingliangyou 268, was used as test materials, by combining controlled irrigation and aeration technology. Four treatments were set up, namely mechanically controlled irrigation aeration (JX), super-micro-bubble controlled irrigation aeration (WP), controlled irrigation (CK), and flood irrigation (YS), to identify effects of aeration on root growth characteristics and water use efficiency of super rice under controlled irrigation conditions. The results showed that effective water saving is up to 15.3% in maximum due to aeration under controlled irrigation as compared with food irrigation condition. Additionally, the root growth is promoted, and the dry matter weight of rice roots increased and dry matter weight of both stem and leaf reduced. So the root volume and thickness as well as dry matter weight improved significantly, and the root activity gets enhanced, while the aging of rice roots gets delayed. In terms of yield, controlled irrigation aeration is similar to flood irrigation, but the seed setting rate, 1000-grain weight, and water use efficiency are better than the latter (Liao et al., 2018).

In the middle and lower reaches of the Yangtze river, rainfall is more in summer. The technology of rain catching and controlled irrigation of rice help to save water by raising the water depth of the field after rain, while the soil water content during the rest period is maintained at 70–100% of field capacity. The objectives of this study were to evaluate rice growth, canopy light utilization, and yield of rice under different rain-catching and controlled irrigation modes (T1: light drought and low storage, T2: light drought and high storage), and to find the optimal storage depth after rain for rice. Measurements included the rice plant's height, tiller number, high tiller growth, leaf angle, canopy interception rate, and yield shape. The plot experiment was conducted in 2012 and 2013, using Nanjing 44 (*Oryza sativa* L.) as the test material. The results showed that T1 treatment improved the height of rice plants and the number of effective tillers in the late growth stage. The number of high tillers had a great influence on the total leaf dry quality; while compared with conventional irrigation (CK), the number of high tillers increased by 11.36% and 7.87% in T1 and T2, respectively. The canopy interception rate of T1 above 0 cm was higher than that in T2 and CK and the leaf area index (LAI) was closely related to 0–40 cm of canopy light distribution. The number of grains per panicle in T1 was lower than in CK and T2; however, the number of grains in T1 was less, and the 1000 grain weight was higher. On 63 days and 83 days after transplanting in 2012 and 78 days after transplanting in 2013, the first, second, and third leaf angles of T1 were larger. Rain-catching and controlled irrigation can increase the dry weight and shoot dry weight of rice, and light drought and low storage (T1) conditions are good for maintaining a high yield because of more tiller number, more grains per panicle, and reasonable light distribution (Lu et al., 2018) (Table 7.7).

On an average, dry matter production, yield, and yield-attributing parameters, such as panicle weight, grain weight per panicle, filled grain (%), panicle number m^{-2}, straw weight m^{-2} and grain weight m^{-2} had greater values in 2014 compared to 2013 (Table 7.8). In both the years, the response of the cultivars to SF was almost similar. SF decreased various yield and yield-attributing parameters in all cultivars, but the effect was higher in cultivars with Sub1. The damage was greater in Swarna Sub1 compared to Savitri Sub1. The reduction in grain weight m^{-2} under SF for Swarna in 2013 and 2014 were 26.9 and 27.4% respectively, while the corresponding figures for Swarna Sub1 were 70.1 and 60.1% respectively. Similarly, SF reduced grain weight per m^{-2} in Savitri Sub1 by 47.1 and 50.5% in 2013 and 2014 respectively, as compared to corresponding figures of 32.6 and 31.1% in Savitri. The Sub1-introgression also depressed shoot biomass and other grain yield parameters, like panicle grain weight, panicle number m^{-2} and filled grain% significantly (Kuanar et al., 2019).

Table 7.7: Crop yield under the controllable irrigation regime.

Year	Treatment	Actual Yield (tha^{-1})	Potential Yield (tha^{-1})	Ear Length (cm)
2012	CK	4.79^{ab}	5.12^{ab}	17.80^{c}
	T1	5.16^{a}	5.35^{a}	18.95^{b}
	T2	4.68^{b}	5.05^{b}	19.54^{a}
2013	CK	3.11^{c}	3.56^{b}	17.23^{b}
	T1	4.39^{a}	4.54^{a}	18.61^{a}
	T2	3.73^{b}	4.33^{ab}	18.54^{a}

Note: In the column, averages followed by common letter(s) are not significantly different at a level of $p < 0.05$.

Trait	Environ	RP	DP	NILs mean range	RUs	CV	LSD
GY(qha^{-1})	NS	78.1	81.6	80.9 62.4–89.4	33.12	6.03	1.65
	S	5.5	52.8	54.1 40.13–63.1	12.70	1.40	-

Table 7.8: Comparison of yield and yield attributes of Sub1 rice varieties under submergence.

Genotype	Treat	Flg. Dt.	Mat. Dt.	Pl. Ht. (cm)	Panicle Wt(g)	Grains (g)/ Panicle	FG/ Panicle	FG (%)
Swarna	C	Oct.16	Nov16	115	3.50	3.27	163	80
	SF	Oct19	Nov22	133	3.35	3.10	173	87
SwarnaSub1	C	Oct20	Nov22	111	3,57	3.33	184	91
	SF	Oct25	Dec03	115	3.32	3.04	172	81
Savitri	C	Nov10	Dec12	116	3.74	3.41	170	85
	SF	Nov13	Dec17	127	3.64	3.19	154	79
SavitriSub1	C	Oct30	Nov30	114	4.03	3.90	164	84
	SF	Nov03	Dec07	118	3.84	3.44	180	75

Grain yield m^{-2} of rice varied significantly for different submergence durations. The highest grain yield m^{-2} was found from D0 (0.62 kg) whereas the lowest yield was recorded from D3 (0.38 kg) treatment (Table 7.9). Grain yield per plot is a function of interplay of various yield components, such as the number of productive tillers, grains panicle^{-1} and 1000-grain weight (Hassan et al., 2003). In present experiment variety there was significant effect on grain yield. The V4 (BRRI hybrid dhan1) produced the highest (0.78 kg) grain yield m^{-2} and the lowest grain weight m^{-2} (0.23 kg) and grain yield m^{-2} of rice varied significantly for different submergence duration. The highest grain yield m^{-2} was found from D0 (0.62 kg) whereas the lowest yield was recorded from D3 (0.38 kg) treatment. Grain yield per plot is a function of interplay of various yield components, such as the number of productive tillers, grains panicle^{-1} and 1000-grain weight (Hassan et al., 2003). In present experiment, the variety had significant effect on grain yield. The V4 (BRRI hybrid dhan1) produced the highest (0.78 kg) grain yield m^{-2} and the lowest grain weight m^{-2} (0.23 kg) was found from V3 (BRRI dhan34) (Tables 7.9 and 7.10).

Table 7.9: Effect of submergence on thousand grain weight and yield of rice.

Treatment	Yield/m² (kg)	Per Cent Reduction of Yield
D0	0.62 a	0.00 d
D1	0.60 a	5.23 c
D2	0.50 b	19.02 b
D3	0.38 c	31.03 a
CV (%)	6.88	7.89

D0 = No submergence, D1 = Six days submergence, D2 =Ten days submergence and D3 = Fourteen days submergence

Table 7.10: Effect of submergence on yield of genotypes.

Treatment	Yield/m^2 (kg)	Per Cent Reduction of Yield
V1	0.36 d	3.86 e
V2	0.44 c	6.91 d
V3	0.23 e	30.43 a
V4	0.78 a	6.63 d
V5	0.73 a	9.00 c
V6	0.61 b	26.10 b
CV (%)	6.88	7.89

V1 = BRRI dhan51, V2 = BRRI dhan46, V3 = BRRI dhan34, V4= BRRI hybrid dhan1, V5 = BRRI hybrid dhan2 and V6 = ACI hybrid1

Interaction of submergence and variety significantly affected the grain yield m^{-2} (Table 7.10). The highest (0.85 kg) grain yield was found from the combination of D0V4 and the lowest (0.12 kg) from D3V3 treatment.

7.4 Submergence Effects on Yield of Inbred and Hybrid Rice

Submergence stress can greatly limit grain yield of inbred rice (*Oryza sativa* L.), but the effects of submergence on hybrid rice are unclear. A pot experiment was conducted to clarify the effects of submergence that happens at the vegetative growth stage of two Chinese hybrid rice cultivars. The rice cultivars, Zheyou-18, an indica-japonica hybrid, and Yliangyou-689, and indica hybrid were planted under two treatments, submergence (43 cm depth of tap water for about two weeks) and control (no submergence). Results showed that the grain yield of submerged 'Yliangyou-689' was 539.0 g m^{-2} in 2018 and 614.2 g m^{-2} in 2019; submergence significantly reduced grain yield by 49.6% and 44.2% of the control. The lower level of grain yield was attributed to reduced survival rate, numbers of tillers and spikelets, plant dry weight, and crop growth rate, and excessive elongation of stems due to submergence of plants. The grain yield of submerged 'Zheyou-18' was 466.8 g m^{-2} in 2018 and 376.4 g m^{-2} in 2019 and neither of which were significantly different from that of the control, indicating higher submergence tolerance in 'Zheyou-18' than in 'Yliangyou-689'. Submergence did not affect plant N and K contents, but reduced the plant N uptake rates by 47.8% to 88.7% and reduced the K uptake rates by 53.9% to 89.5% of that of the control in these two hybrid cultivars. Furthermore, insignificant differences were observed in all of the parameters related to rice starch viscosity between control and submergence treatment (Chen et al., 2021) (Table 7.11).

Table 7.11: Grain yield and yield components of two rice cultivars subjected to submergence or not (control).

Year	Cultivars	Treatments	Grain yield gm^{-2}	Spikelets number m^{-2}	Panicle number m^2	Spikelets per panicle	Grain filling %	Grain weight g
2018	Yliangyou-689	Control	1069.9a	52708a	296.1a	182.2a	77.9a	22.4a
		Submergence	539.0b	34145b	133.8b	256.7a	66.7b	20.4b
	Zheyou-18	Control	661.9a	42211a	191.0a	220.2a	67.1b	20.2a
		Submergence	466.8a	20998b	148.0a	144.3b	99.5a	21.7a
2019	Yliangyou-689	Control	1101.8a	66571a	390.4a	169.7a	70.9a	20.2a
		Submergence	614.2b	34886b	261.9b	187.4a	72.7a	20.9a
	Zheyou-18	Control	521.5a	55205a	223.9a	249.8a	40.8a	20.0a
		Submergence	376.4a	44524a	161.9a	275.4a	37.5a	19.3a
		ANOVA results						
		Year (Y)	NS	##	##	NS	##	NS
		Treatment (T)	##	##	##	NS	NS	NS
		Cultivar (C)	##	NS	##	NS	NS	NS
		Y × T	NS	NS	NS	NS	NS	NS
		C × T	##	NS	NS	NS	NS	NS

Within columns for each treatment, means followed by different letters are significantly different according to LSD ($P < 0.05$).

#,## Significant at the 0.05 and 0.01 probability levels, respectively, NS: Nosignificant.

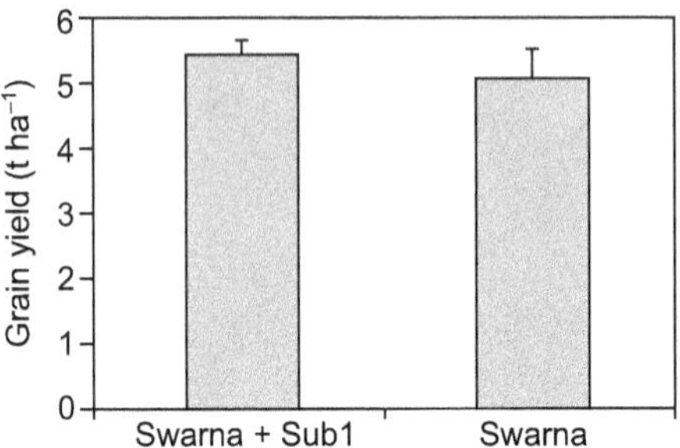

Fig. 7.1: Grain yield of Swarna + Sub1 and Swarna under favorable rain-fed lowland conditions. Data are means of four replications and vertical bars represent standard deviation.

Application of 30 lb/acre N as compost or $(NH_4)_2SO_4$ to waterlogged rice increased grain yields by 45 %. Simultaneous application of 20 lb K_2O and/or 20 lb P_2O_5 did not materially affect yields, due to the generally high exchangeable K content in waterlogged soils and to the low P requirement of the crop.

Through marker-assisted breeding, Sub1 was also introgressed in Swarna, a popular rice cultivar in South and Southeast Asia and is currently being tested in farmers' fields in India and Bangladesh. Introgression of Sub1 into Swarna greatly enhanced its survival under submergence, and with no other obvious effects on original characteristics of Swarna. Preliminary field experiments under normal conditions showed that both cultivars have similar grain yield (Fig. 7.1). The dramatic effect of Sub1 on what is essentially a quantitative trait, suggests a regulatory locus rather than a specific enzyme. Efforts are currently in progress to introduce Sub1 into popular high-yielding rice varieties of rain-fed lowlands. This approach will substantially enhance submergence tolerance of these varieties without sacrificing their quality aspects or yield potential, and will dramatically shorten the breeding cycle. Another level in the hierarchy of genetic controls which have a bearing in regulating stress responses is exercised at the signal transduction pathway (Sarkar et al., 2006).

With no other obvious effects on original characteristics of Swarna, preliminary field experiments under normal conditions showed that both cultivars have similar grain yield (Fig. 7.1).

References

Chen, Yutiao, Song, Jiayu, Yan, Chuan, and Hong, Xiaofu. (2021). Effects of submergence stress at the vegetative growth stage on hybrid rice growth and grain yield in China. *Chilean Journal of Agricultural Research*, 81(2): 191–201. https://dx.doi.org/10.4067/S0718-58392021000200191.

Das, K.K., Panda, D., Sarkar, R.K., Reddy, J.N., and Ismail, A.M. (2009). Submergence tolerance in relation to variable flood-water conditions in rice. *Environ. & Exp. Bot.*, 66: 425–434.

Devkota, K.P., Koichi, F., Valere, C.M., and Humpherys, E. (2022). Does wet seeding combined with Sub1varieties increase yield in submergence prone lowlands of West Africa? *Field Crops Research*, 276.

Dwivedi, S.K., Bhakta, N., Kumar, S., and Misra, J.S. (2017). Characterization of Submerged Tolerant Elite Rice Genotypes Having Improved Physiological Traits and Oxidative Defence System grown under Rain-fed Lowland Ecosystem of Eastern Indo-Gangetic Plains. *Agric. Res.*, 6: 207–213.

FAO. (1999). *Report of the Expert Consultation (Dr. D.V. Tran) on Bridging he Rice-yield Gap in the Asia-Pacific Region*, Bangkok, Thailand, December 1999.

Irmawati, Syawal, Y., Sulistyaningsih, L.N., Susilwati, S., Yakup, Y., and Ronaldo, E. (2020). The evaluation on recovery phase of post submerged rice. *Russian J. Agric and Socio-economic Sciences*,108: 159–166.

Ismail, A.M., Singh, U.S., Singh, S., Dar, M.H., and Mackill, D.J. (2013). The contribution of submergence–tolerant (Sub1) rice varieties to food security in flood-prone rain-fed lowland areas in Asia. *Field Crops Res.*, 152: 83–93.

Jin-hu, N. (2014). Effects of waterlogging stress on rice morphology and yield component at the jointing stage. *Chinese Journal of Ecology*.

Kuanar, S.R., Molla, K.A., Chattopadhyay, K., Sarkar, R.K. and Mohapatra, P.K., et al. (2019). Introgression of *Sub1 (SUB1) QTL* in mega rice cultivars increases ethylene production to the detriment of grain-filling under stagnant flooding. *Sci. Rep.*, 9: 18567.

Li, C., Shi, Y.Y., and Chen, Y.Y. (2004). On influence of flood and waterlogging on yield reduction of rice. *Journal of Natural Disasters*, 23: 83–87 (in Chinese with an abstract in English).

Liao, J., Hu, D., Pei, Y., Chen, Z., Ding, X., and Luo, D. (2018). Effects of aeration on root growth and water use efficiency of super rice under controlled irrigation conditions. *Journal of Drainage and Irrigation Machinery Engineering*, 36: 920–924.

Lu, H., Qi, X., Guo, X., Towa, J.J., Zhen, B., Qiao, D. et al. (2018). Canopy Light Utilization and Yield of Rice under Rain-Catching and Controlled Irrigation. *Water*, 10: 1340.

Liu, X., Lu, H., Yang, S., and Wang, Y. (2016). Impacts of biochar addition on rice yield and soil properties in a cold waterlogged paddy for two crop seasons. *Field Crops Research*, 191. Doi: 10.1016/j.fcr.2016.03.003.

Mahapatra, N. (2017). Yield and its attributing characters of different rice genotypes to submergence stress. *The Pharma Innovation Journal*, 6: 315–319.

Sarkar, R., Reddy, J.N., Sharma, S., and Ismail, A.M. (2006). Physiological basis of submergence tolerance in rice and implications for crop improvement. *Curr. Sci.*, 91: 899–906.

Shan, L., Deng, X.P., Su, P., Zhang, S.Q., Huang, Z.B., and Zhang, Z.B. (2000). Exploitation of crop drought resistance and water-saving Potentials – Adaptability of the crops to the low and variable water conditions. *Review of China Agricultural Science and Technology*, 2: 66–70 (in Chinese).

Sharma, A., and Jaiswal, H. (2021). A study of gene action for yield and quality traits of Basmati rice (*Oryza sativa* L.) following Hayman's graphical approach. *Oryza*, 58: 8–14.

Shenoy, H. (2020). Effect of Neem cake urea mixed application on growth and yield of rice under submerged condition. *Int. J. Curr. Microbiol. App. Sci.*, 9(04): 2224–2229.

Singh, A., Singh, A.K., Nehal, N., and Sharma, N. (2020). Performance of different zinc sulphate level on survival, morphological and biochemical parameters of Sub1 and Non Sub1 rice varieties under submerged condition. *J. Pharmacognosy and Phytochemistry*, 9: 2505–2512.

Sultana, T., Ahamed, K.U., Naher, N., Islam, M.S., and Jaman, M.S. (2018). Growth and yield response of some genotype under different duration of complete submergence. *JAERI*, 15: 1–11.

Turner, N.C. (1997). Further progress in crop water relations. *Advances in Agronomy*, 58: 293–338.

Wang, B., Zhou, Y.J., Xu, Y.Z., Chen, G., Hu, Q.F., Wu, W.G., et al. (2014). Effects of Waterlogging Stress on Growth and Yield of Middle Season Rice at the Tillering Stage. *China Rice*, 20: 68–72, 75.

Yadav, G.S., Subhash Babu, Das, A.K., Bhowmik, S.N., Dutta, M., and Singh, R.P. (2019). Soil carbon dynamics and productivity of rice-rice system under conservation tillage in submerged and unsubmerged ecologies of eastern Indian Himalaya. *Carbon Management*, 10: 51–62.

Yamungmorn, S., Rinsinjoy, R., Lordkaew, S., Dell, B., and Prom-U-Thai, C. (2020). Responses of grain yield and nutrient content to combined zinc and nitrogen fertilizer in upland and wetland rice varieties grown in waterlogged and well-drained condition. *J. Soil Sci. Plant Nutr.*, 20: 2112–2122.

Index

U

V

W

X

Y

Z

www.ingramcontent.com/pod-product-compliance
Lightning Source LLC
Chambersburg PA
CBHW060405220326
41598CB00023B/3029
* 9 7 8 1 0 3 2 6 6 3 4 0 1 *